The Institute of Biology's
Studies in Biology no. 111

Microorganisms and Man

W. C. Noble
Ph.D., F.I.Biol., M.R.C.Path.
Reader in Bacteriology,
Institute of Dermatology,
London

Jay Naidoo
B.Sc.
Research Assistant,
Department of Bacteriology,
Institute of Dermatology,
London

Edward Arnold

© W. C. Noble and Jay Naidoo, 1979

First published 1979
by Edward Arnold (Publishers) Limited
41 Bedford Square, London WC1B 3DQ

Paper edition ISBN: 0 7131 2746 5

All Rights Reserved. No part of this publication may be reproduced, stored in a retrieval system, or transmitted, in any form or by any means, electronic, mechanical, photocopying, recording or otherwise, without the prior permission of Edward Arnold (Publishers) Limited.

Printed in Great Britain by
Thomson Litho Ltd, East Kilbride, Scotland

General Preface to the Series

Because it is no longer possible for one textbook to cover the whole field of biology while remaining sufficiently up to date, the Institute of Biology has sponsored this series so that teachers and students can learn about significant developments. The enthusiastic acceptance of 'Studies in Biology' shows that the books are providing authoritative views of biological topics.

The features of the series include the attention given to methods, the selected list of books for further reading and, wherever possible, suggestions for practical work.

Readers' comments will be welcomed by the Education Officer of the Institute.

1979   Institute of Biology
                                              41 Queen's Gate
                                              London SW7 5HU

Preface

In this short book we have tried to show some of the many ways in which microorganisms affect our lives. Microbial action on various, sometimes similar, substrates can be seen to yield substances useful or even vital to man, or to result in destruction and decay. The student can keep up to date on many of the facets of microbiology to which he is introduced in this book by means of the daily newspapers. Food poisoning will continue to occur on cruise ships and at weddings. Rabies and Dutch elm disease will reappear as newsworthy topics, as will the need for single-cell protein and disposal of industrial waste. We hope that, in this way, microbiology will be seen as a living scientific discipline of interest and importance to all.

London, 1978   W.C.N.
                                                            J.N.

Contents

General Preface to the Series   iii

Preface   iii

1 Introduction   1

2 Microbial Production Processes   4
2.1 Milk and cheese products   2.2 Production of alcohol   2.3 Vinegar production   2.4 Other food production processes   2.5 Yeasts as food   2.6 Production of antibiotics

3 Some Results of Microbial Degradation   24
3.1 Microbial activity in sewage disposal   3.2 Wash-day enzymes   3.3 Biodeterioration of materials   3.4 Biodeterioration of pollutants   3.5 Microbial deterioration of food

4 Diseases of Man and Animals   45
4.1 Food poisoning and other gastrointestinal diseases   4.2 Aerosol infections   4.3 Contact infections   4.4 Vector-borne diseases

5 Diseases of Plants   58
5.1 Mechanisms of plant disease   5.2 Transmission of plant pathogens   5.3 Viral plant diseases

6 Postscript   72

Suggestions for Further Study   74

1 Introduction

Microorganisms are usually thought of solely in terms of the damage they cause to humans, farm animals and crops, or to buildings, clothes and food. This damage we call disease and decay. Yet without microorganisms, life as we know it would be impossible: cows could not digest grass without a rich rumen flora to break down the cellulose; oil would not exist; the world would be cluttered with undecayed leaves. The formation of compost and return of nutrients from dead plants and animals to the soil is one aspect of 'decay'–the decomposition on which man depends. There is another aspect, however, and that is the use that man makes of the synthetic and degradative abilities of microorganisms to provide useful chemicals. For example, in each year during the mid-1970s some 600 000 million hectolitres of beer worth about £15 000 million were produced in the world and industrial alcohols (methylated spirit) reached 20 million tonnes production with a value of £350 million. Fermentation processes produced 265 000 tonnes of citric acid (£160 million) and 200 000 tonnes of L-glutamic acid (£150 million). The pharmaceutical industry annually produced a total value of £550 million of antibiotics including 7000 tonnes of penicillin (£112 million) and 3500 tonnes of tetracycline (£45 million). With very many other products the world annual value of synthetic microbial products was about £36 600 million.

All microorganisms have some characteristics in common. The principal one which concerns us is that it is *through the cell wall* that all nutrient materials are absorbed and all excess or 'waste' products are excreted. Some enzymes are produced in considerable quantities extracellularly, but this can be seen as a necessity for an organism which utilizes, for example, starch or cellulose, for these substrates must first be degraded to a form in which they can be absorbed through the cell wall before further breakdown can take place. In some cases secondary metabolites are excreted in vast amounts, as is the case with the antibiotics. Sometimes the products are harmful to man as in the exotoxins which cause food poisoning.

There are three principle forms of microorganism described in this book: fungi, bacteria and viruses. The fungi can be divided into two groups: (*a*) filamentous fungi, which form a thread-like mycelium 3–5 μm wide, have specialized spore bodies, and include moulds such as *Penicillium* and *Aspergillus*; and (*b*) yeasts, such as *Saccharomyces*,

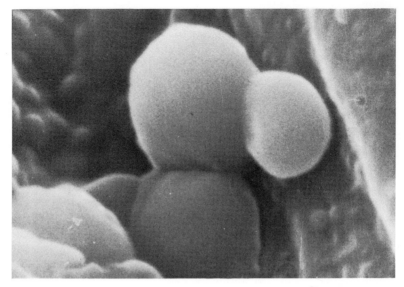

Fig. 1-1 Scanning electron microscope picture of yeasts on human skin. Note the budding which is characteristic of yeasts.

which are best known as forming single cells 3–5 µm in diameter and which reproduce by budding, although pseudomycelium may be formed. Bacteria reproduce by dividing, not by budding, and like the fungi are mostly free-living and able to reproduce without the aid of another living cell. Many bacteria are less than 2 µm wide, though some may be up to 10 µm long; many are spherical or form only short rods. In the laboratory they are usually divided according to their shape and to their reaction to Gram's stain. *Lactobacillus*, *Streptococcus* and *Pseudomonas* are amongst the bacteria referred to in the following pages. Virus nucleic acid does not code for all essential enzymes, viruses are therefore obligate intracellular parasites and usually damage the host cell. They are chemically amongst the simplest of the microorganisms; tobacco mosaic virus, for example, can be crystallized yet retains its infectivity. One special group of viruses, known as phage, infect bacteria (all viruses are very small, up to 100 phages can be found inside a single bacterial cell 1 µm in diameter). Under optimal conditions, phages can reproduce in less than 30 minutes releasing 50–100 new phage particles to infect more bacteria. Phage infection of cultures used in an industrial process can therefore be detrimental to the end-product. The wide variety of structure and the behaviour of viruses in their host cells are described in HORNE (1978).

The next two chapters describe a variety of industrial processes which use microorganisms. It might seem that the production of cheddar cheese has little in common with, for example, the purification of water, yet frequently the chemical reactions carried out by the microorganisms are very similar. The end product is different only because care is taken to limit the degree of change in cheese production whilst in sewage disposal reactions are allowed to go to completion. It is convenient, but artificial, to divide microbial processes into production and destruction since alcohol 'production' could equally be described as 'biodegradation of sugar'.

The last two chapters are concerned with disease in animals and in plants. An organism that causes disease is known as a pathogen. The examples have been chosen to illustrate the variety of routes by which a pathogen can enter a host: by ingestion in contaminated food or drink; by inhalation of, or contact with, airborne particles; by direct contact; by inoculation via an animal bite; or by skin puncture by an arthropod vector. There are many similarities in the manner of spread of pathogens in groups of plants and in groups of animals, including man.

2 Microbial Production Processes

2.1 Milk and cheese products

An examination of the microbiology of cheese and yoghurt shows how a range of microbes (usually bacteria) act on a common substrate, lactose, to yield a number of different products. Milk is a complex emulsion containing protein, fat, sugar and minerals. Cow's milk is about 87% water, 3.5% protein, 3.5% fat and 5% lactose. These figures must necessarily be approximate since not only do breeds differ (Friesian cattle produce the most milk, but it has on average only 3.4% fat, whilst Jerseys produce the least milk, but it has 5.1% fat), but also the milk from each of the four teats on the udder can be slightly different in composition. Lactose is the main sugar in milk, it is a disaccharide and is hydrolysed to give equal amounts of glucose and galactose.

About 80% of milk protein is casein, present as a colloidal suspension in complex particles containing calcium and phosphorus. At a pH of 5.2–5.3 precipitation of casein occurs and, at pH 4.6–4.7 (the iso-electric point), casein is least soluble and is free of all inorganic salts. Casein can be coagulated by rennet, but it is lactic acid derived from fermented lactose which usually coagulates milk during cheese production.

2.1.1 Fermented milk products

Microbiology. The chief microorganisms used in cheese production are species of *Lactobacillus* and *Streptococcus*. Both are Gram-positive, unicellular, free-living, non-sporing and non-photosynthetic organisms; *Lactobacillus* sp. forming rods less than 2 μm wide whilst the streptococci, also less than 2 μm wide, form spheres in a chain formation.

Yoghurt. *Lactobacillus bulgaricus* and *Streptococcus thermophilus* in equal amounts are used in the production of yoghurt. In a typical process milk is pasteurized and homogenized and then, at a temperature of around 45°C, about 2.5% of starter culture is added. It is packed into cartons and fermented in a hot-room at 45°C for about four hours, after this time the pH will have dropped to 4.6 and the milk solids will have coagulated. The fermentation reaction can be summarized as follows:

$$\underset{\text{glucose}}{C_6H_{12}O_6} \rightarrow \underset{\text{lactic acid}}{2CH_3CHOHCOOH}$$

the glucose being derived from lactose in the milk. The reaction is, of course, more complex than shown here and for streptococci follows the Embden–Meyerhof pathway. After cooling, plain yoghurt will keep for approximately three weeks, but the presence of whole fruit, which is added before fermentation, reduces the shelf-life.

Cottage cheese. A similar process is used with cottage cheese, though there are important differences. The starter cultures are *Streptococcus lactis* and *S. cremoris* and the incubation takes place over six hours at 31°C. A pH of 4.6 is reached, again due to the production of lactic acid, and coagulation of milk protein follows. This curd is then 'cooked' at 49°C, the whey is drained off (Fig. 2-1) and salt is added before the curd is finally pressed. Cottage cheese should be eaten within about two weeks of its production.

Cheddar cheese. Rennet is used to help produce the initial coagulation in cheddar cheese. Starter cultures of *S. thermophilus* are added to pasteurized milk and the addition of rennet produces the curd. After cooking at 38°C for 45 minutes the 'cheddaring' process

Fig. 2-1 Whey being drained from curds during cheese-making. (This picture is taken from a slide/tape presentation by courtesy of Camera Talks Ltd. The original was provided by Unigate Ltd.)

takes place. Cheddaring originated from the English village of Cheddar and was brought into standard commercial practice in 1875. Originally, raw unpasteurized milk was used for cheese-making and, because of the presence of bovine faecal organisms (especially the Gram-negative bacilli *Escherichia* and *Aerobacter*), fermentation of the lactose could result in the production of excessive amounts of carbon dioxide, hydrogen and hydrogen sulphide, giving a nasty smell and unacceptable taste to the cheese. The cheddaring process consists of cutting the warm curd into blocks and stacking these on each other. The pH rapidly falls to 5.2–5.3 and the coliforms are killed or inhibited. Piling up also flattens out gas pockets and drives gas out of the curd. Cheddaring is now used all over the world so that, for example, New Zealand 'cheddar' and American 'cheddar' can be bought.

The choice of 38°C for cooking governs the choice of starter culture. Although 38°C does not kill *S. lactis* and *S. cremoris*, it heat-shocks them so that further development is slow. Temperatures up to 43°C are possible if *S. thermophilus* (the 'heat-lover') is used, and up to 49°C if *S. faecalis* or *S. durans* are used. Bacterial enzymes active in cheddar are responsible for the flavour and aroma which develop during ripening or maturation: protease, peptidase, lipase, amino acid decarboxylase and deaminase all contribute to the formation of ammonia, acetic acid, alcohol, diacetyl, acetyl methyl carbinol, butyric acid, caproic acid, secondary butyl alcohol and methyl ethyl ketones. Many of these compounds are derived from the fermentation of glucose.

2.1.2 Microbial ripening of cheese

Reference was made above to the ripening or maturing process which cheddar undergoes as a result of the continued action of the starter cultures. Some cheeses have specific bacterial or fungal ripeners added and these greatly modify the appearance and taste of the final product.

Limburger cheese, Bel Paese cheese. These, and other similar cheeses, are formed by a rennet coagulation of sweet pasteurized milk followed by the addition of a lactic starter, a light cooking and salting. After the cheese has been formed it is 'smeared' with *Brevibacterium linens*, a Gram-positive rod, which produces a reddish brown surface growth. The ripening shelves are heavily contaminated with *B. linens* and smearing is needed only to ensure an even coat. This bacterium produces aromatic compounds which contribute significantly to the final flavour.

Gorgonzola, Stilton and Roquefort cheeses. All cheeses in this group are ripened by the growth of moulds in 'veins'. Spores of *Penicillium roqueforti* or *P. glaucum* are added to the lactic acid and rennet-induced curds. Natural cavities in the cheese become lined with green spore-bearing mycelium. Stilton has to be holed with needles to introduce air so that the aerobic moulds can grow.

Camembert cheese. *P. camemberti* is a white-spored mould and is coated onto the outside of the cheese. It may fail to grow, however, unless surface yeasts and other organisms have restored the pH to 7.2 following lactic acid production and curd formation.

Brie. This cheese has the distinction of being ripened by both *B. linens* and *P. camemberti*.

2.2 Production of alcohol

Alcohol is another end product of carbohydrate metabolism, and is usually the result of anaerobic fermentation by yeasts. The microbial production of alcohol can be summed up in the equation:

$$C_6H_{12}O_6 \to 2C_2H_5OH + 2CO_2$$

sugar → alcohol + carbon dioxide

This suggests that slightly over half the weight of sugar should be converted into alcohol, but other fermentation products usually account for a slight 'loss' so that about 90% of the expected yield is obtained. These other products may be valuable in providing additional taste and aroma where the alcohol is destined for consumption, but are merely 'impurities' in industrial processes.

Alcohol has many uses; the precise use governs the choice of raw material and hence the microbiological processes. It can be used as a car fuel when mixed with other ingredients and also has a great part to play as an industrial solvent. Much of this industrial alcohol is made by chemical synthetic processes from ethylene or acetylene. Up to one third may be produced by microbial fermentation, the precise amount depending on the availability and cost of raw material (which may be, for example, surplus molasses production in Cuba and the West Indies).

The key to the production of alcohol in wines and beer and for spirit distillation, lies in the yeast species and also in the source of sugar fermented. In wine production the monosaccharide glucose, which is naturally present in the grape juice, is fermented directly to alcohol. Occasionally the disaccharide sucrose is added; this can be hydrolysed by the yeasts to give the monosaccharides glucose and fructose. In beer-making, the polysaccharide starch is first broken down to the disaccharide maltose by enzymes in the grain and then

8 PRODUCTION OF ALCOHOL §2.2

further broken down to glucose, for the yeasts are unable to attack maltose directly. Reference has already been made to the disaccharide lactose which is found only in milk; this is not attacked by *Saccharomyces cerevisiae* so it can be added as a sweetener, for example in 'milk stout'. *Streptococcus lactis* and *Strep. fragilis*, however, can ferment lactose, so milk can be used to produce an alcoholic drink as in 'koumiss', made from asses milk.

Cane sugar and beet sugar are chemically identical (though many wine producers will only use cane sugar). Both are the disaccharide sucrose which is hydrolysed, by enzymes known as invertases, to the monosaccharides glucose and fructose:

$$C_{12}H_{22}O_{11} + H_2O \rightarrow C_6H_{12}O_6 + C_6H_{12}O_6$$
sucrose + water → glucose + fructose

Glucose and fructose have the same general formula and are said to be isomeric. (Isomers differ from each other much as a pair of gloves differ. It was a study of isomers that started Pasteur on the road to microbiology.) The mixture of glucose and fructose derived from sucrose is called 'invert sugar' and is often used in the food industry.

Microbiology. The principle yeasts in alcohol production are the *Saccharomyces*. (Recent studies have suggested changes in the nomenclature of the *Saccharomyces* but these have not yet been fully accepted by the brewers and vintners and in this book the names still in wide use in the industry are used.) *Saccharomyces* are spherical, ellipsoidal or occasionally cylindrical yeasts which reproduce vegetatively by multilateral budding and rarely form a pseudomycelium.

S. cerevisiae is used for 'top fermented' British beers whilst *S. carlsbergensis*, a 'bottom fermenter', is, as its name suggests, used in lager production. *S. cerevisiae* var *ellipsoideus* (*S. ellipsoideus*), which has a high alcohol tolerance, is used for wine production. Typical of the 'sherry yeasts', which also tolerate high alcohol levels, is *S. fermentati*. A strain of this yeast, sometimes known as *S. beticus*, forms a skin, or pellicle, on the surface of the developing wine. Since this skin excludes oxygen from the fluid below, the dark-coloured oxidation products are not formed. The result is pale sherries, 'finos' that are less sweet than the dark 'oloroso' type.

During aerobic respiration, yeasts are able to oxidize glucose to CO_2 and water with the release of considerable amounts of energy available for metabolism:

$$C_6H_{12}O_6 + 6O_2 \rightarrow 6CO_2 + 6H_2O + 2929 \text{ KJ}$$
glucose + oxygen → carbon + water + energy
                      dioxide

When high yields of yeast are required, as in production for the baking industry, cultures are highly aerated and an excess of sugar is avoided; in this way maximal yields of yeast are obtained and little or no alcohol is formed. However, when O_2 is limited, the yeasts act as facultative anaerobes and fermentation, rather than oxidation, takes place releasing a smaller amount of energy and leaving alcohol as an end product:

$$C_6H_{12}O_6 \rightarrow 2C_2H_5OH + 2CO_2 + 276\,KJ$$

glucose → alcohol + carbon dioxide + energy

2.2.1 Sources of drinkable alcohol

The main sources of fermentable material for alcohol production are naturally-occurring sugars, starch or cellulose.

Wine production. Alcohol in wines is derived from sugar in the grapes, although for certain types of wine (or in poor harvest years) sucrose from cane sugar may be added. The yeasts occur naturally on the grape skins or may be laboratory-derived cultures of *S. cerevisiae*.

Beer production. Barley is the starch source for whisky and beer. The grain is germinated when enzymes of the diastase complex convert starch initially into dextrins and maltose (α amylase); subsequently the dextrins are also converted to maltose by β amylase. In a typical brewery process the grain is first 'malted': it is soaked in water cisterns for 48–72 hours, where it absorbs about its own weight of water; it is then spread on concrete floors to germinate for eight to twelve days. From time to time the grain is turned over to ensure even germination. During this malting period the diastase enzymes begin to convert the starch to sugar. Germinated grain is then dried to stop growth of the root and shoot so that it can be stored; drying the grain prevents spoilage by moulds. This process is known as 'kilning'. The degree of kilning affects the end product, grain destined for dark beers and stout is kilned at a higher temperature than that for pale ales. The diastase enzymes are allowed to complete their work during 'mashing', for only about 10% of the starch is converted during malting. The dried, crushed, germinated grain is mixed with water in tanks called 'mashing tuns' and heated to the optimum temperature for enzyme action. Alpha amylase has an optimum temperature of 67°C and an optimum pH of 5.7; β amylase has optima of 54°C and pH 4.7. During mashing the remaining starch is converted to sugar. Yeasts are then added to convert the sugar to alcohol. The processes differ in principle and practice from this point for the brewer is interested in the formation of flavours as well as alcohol.

Beer is brewed in immense quantities and 450 000-litre tanks of mash liquor need large quantities of starter yeast. Accordingly, to ensure that he has sufficient yeast, the brewer must organize a series of cultures of increasing volume starting with a test-tube in the laboratory and ending with a starter culture of several hundred litres. These cultures must not become grossly contaminated or a complete batch of beer may be lost. (This is not simply a loss of so many bottles of saleable beer, but poses disposal problems for it cannot simply be poured down the drain – the impact on the microflora of a sewage works of a vast quantity of spoiled beer could be disastrous, leading to the breakdown of the works.)

Although brewing is traditionally a batch process, continuous fermentation systems are in use. These consist of two fermentation tanks; sterilized 'wort', the malt infusion derived from the mash, is pumped into the first tank where yeast growth occurs with about half the desired production of alcohol. From the first tank the wort flows to a second in which more alcohol is produced but little growth of the yeast occurs; this control is achieved by careful regulation of the oxygenation. The final stage consists of a conical tank fitted with cooling coils. Cooling helps the yeast to settle out and it is drawn from the bottom of the tank, CO_2 is drawn from the top of the tank and beer from the middle. The economic advantages of the method include some saving on labour costs, minimal losses of beer, retention of flavours which may otherwise be lost in foaming, and easy collection of CO_2 (a valuable by-product sold in the solid form as dry-ice for cooling). CO_2 is also put back into some beers under pressure before sale to achieve a good 'head', for many modern beers are pasteurized to extend shelf-life.

Spirit production. It is interesting to contrast beer production with the fermentation of grain or potato for gin production and with the fermentation of grapes to form some of the world's finest wines. Pure alcohol is required for gin and vodka manufacture, for the product is either flavoured later, as in the case of gin, or marketed as a flavourless drink. Modern techniques can be summarized as follows. Potatoes, or grain such as maize, are first cooked in a continuous process. Milled grain, malt and hot water are mixed to form a slurry and pumped into a wide-bore stainless steel tube injected with steam. Enzymes in the malt serve to start early liquifaction of the starch and improve its handling properties. During cooking, a temperature of about 150°C at 4.5×10^5 Pascals (65 lb/in^2) is held for about ten minutes as the slurry passes down the tube. The slurry is then cooled to 63°C and further malt added; conversion of starch to sugar takes place within a few minutes and after further cooling to 27°C, the slurry is discharged into the fermenter. Between 0.1 and 0.15 kg of yeast are needed for each 1000 litres of mash in the fermenter. *Saccharomyces cerevisiae* strains adapted to the particular environment of the fermenter are used, being maintained in pure culture. From test-tube culture to fermentation mash requires at least three sub-cultures of increasing bulk in stainless steel vessels which are sterilized after use, although it may be possible to keep cultures in the yeast plant for up to one month without renewal from the laboratory source.

2.2.2 Sources of industrial alcohol

Starch can be hydrolysed by dilute mineral acids, such as hydrochloric acid or sulphuric acid, the mixture being neutralized with calcium carbonate (chalk), sodium carbonate, sodium hydroxide or ammonia before fermentation. Wood, in waste forms such as sawdust, can also be converted to sugar by acid hydrolysis, though this may require 40% hydrochloric acid. The spent sulphite liquor from paper mills contains about 2% sugars derived from the wood chips and this too can be fermented to alcohol in a commercially profitable process. Wood products as a source of industrial alcohol are common in Scandinavia, but in Britain molasses, either imported or derived from sugar refining, is used as the principal source. Molasses is about 50% sugar (two thirds sucrose and one third invert sugar); it is diluted to about 15% sugar. In this way approximately 7.5% of alcohol is achieved in two or three days, which is more economic than trying to achieve a higher percentage of alcohol over a longer period. Beet molasses contains sufficient nutrient to sustain yeast growth, but ammonium sulphate and phosphate must be added to cane molasses. The pH is brought to 4.5 to encourage glycolytic enzymes and discourage unwanted bacteria. In the U.S.A., where grain is the usual ingredient for industrial alcohol, the pH was formerly adjusted by a culture of *Lactobacillus delbruckii*, to produce lactic acid, a process known as lactic souring.

During distillation for either drinkable or industrial alcohol the main contaminants are alcohols other than ethanol. In the simple 'pot-still' used for whisky, brandy and rum, methyl alcohol distils over first. Amyl alcohol, the main ingredient of fusel oil, distils off after the ethyl alcohol. These 'heads' and 'tails' are redistilled and eventually discarded. The pot still is an inefficient device which permits other substances known as 'congenerics' to distil over with the ethyl alcohol. These congenerics consist of esters, organic acids and other alcohols and are vitally important to the taste and aroma of the final product. In industrial processes, however, a pure, or at least 95% pure, alcohol is needed and a 'compound-still' consisting of an 'analyser' and a 'rectifier' are used. This latter is analogous to the packed plate, bellhood, or bubble-fractionating column and enables the distiller to select certain fractions or 'cuts' of the distillate, thus ensuring a constant product.

2.3 Vinegar production

The chemistry of the process can be summarized as follows:

$$C_6H_{12}O_6 \rightarrow 2C_2H_5OH + 2CO_2$$
$$\text{glucose} \rightarrow \text{alcohol} + \text{carbon dioxide}$$

$$C_2H_5OH + O_2 \rightarrow CH_3COOH + H_2O$$
$$\text{alcohol} + \text{oxygen} \rightarrow \text{acetic acid} + \text{water}$$

The first reaction may be carried out anaerobically by yeasts such as *Saccharomyces cerevisiae* var *ellipsoideus*. (In practice there are many side reactions and the production of alcohol occurs through a number of small steps.) The second reaction is aerobic and is brought about by bacteria of the *Acetobacter* group. Different species occur in different processes. The conversion of alcohol to acetic acid occurs via acetaldehyde and hydrated acetaldehyde:

$$CH_3CH_2OH + O_2 \rightarrow 2CH_3CHO + 2H_2O$$
$$\text{acetaldehyde}$$

$$CH_3CHO + H_2O \rightarrow CH_3-\underset{OH}{\overset{H}{C}}-OH$$
$$\text{hydrated acetaldehyde}$$

$$2CH_3-\underset{OH}{\overset{H}{C}}-OH + O_2 \rightarrow 2CH_3COOH + 2H_2O$$

Microbiology. *Acetobacter* are involved in vinegar production and are Gram-negative, straight, rod-shaped organisms, less than 2 μm wide. They are aerobic and some species are motile by means of flagellae. The present tendency is to refer all strains of acetic acid bacteria to the genus *Acetobacter* and some workers restrict the species to *A. aceti*, *A. curvum*, *A. scheutzenbachii*, *A. acidophilum* and *A. rancens*. All other previously described species are regarded as varieties of these.

Production. There are two basic processes of vinegar production—'slow' and 'quick'. The *'slow'* method utilizes spontaneous alcoholic fermentation to about 11–13% alcohol by yeasts present on the grape to form a low-grade wine. This alcoholic solution is then allowed to undergo acetic acid fermentation in partially filled barrels by naturally occurring species such as *A. scheutzenbachii*, *A. curvum* and *A. orleanense*. A variant on this method uses a continuous process in which 25–30% of the charge consists of raw vinegar providing a ready source of acetic acid bacteria (Orleans process). This process may still require weeks or months for completion, but it produces a less harsh vinegar than the quick methods for there is some maturation of the product in the barrel.

'*Quick*' *methods* rely on actively moving the fluid during processing, usually over large surface areas, so that good oxygenation occurs; the starting product is an alcoholic liquid produced naturally or, in the U.S.A., produced by synthesis of alcohol from ethylene.

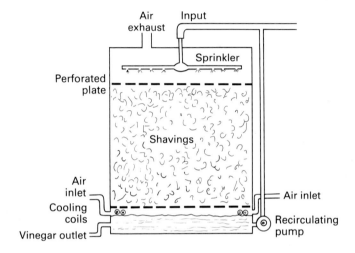

Fig. 2-2 Diagrammatic version of 'Fring's acetator', acetic acid producer.

An example of a 'quick' method is Fring's generator or acetator (Fig. 2-2). The wooden cylindrical tank is built of three sections, the centre one being packed with beechwood shavings, corn cobs, charcoal, coke, or some other material that will yield a large surface area without packing solid. The top section contains a spray mechanism from which the alcoholic solution is sprinkled onto the packing. As it passes over the packing, which acquires a heavy growth of acetic acid bacteria, the alcohol is converted to acetic acid. In the bottom section the fluid is collected and may be passed again through the same or a similar generator until an acceptable degree of acetic acid production has occurred and little alcohol remains. Considerable heat may be generated by the process and this is taken away by the air which circulates through the column or, in some systems, by cooling coils at the base of the column. *A. aceti*, *A. rancens* and *A. xylinum* may be active in these generators, *A. orleanense* may also be found. Mixed cultures are usually more effective than single cultures.

Submerged cultures are being increasingly used. A stirred medium containing 8–12% alcohol is inoculated with *A. acetigenum* and held at 24–29°C. Aeration is achieved mechanically by blowing in air through fine jets. Cooling coils are necessary in such a fermenter to keep the temperature below about 35°C. Conversion of alcohol to acetic acid is about 98% efficient however.

In one modern version of the submerged culture technique (patented in 1972) a tower, 60 cm wide and up to 6 m high, made of polypropylene reinforced with fibreglass, is used. Aeration is carried out through a plastic perforated plate at the bottom of the tower. The fluid is supported on this perforated plate. Pilot-scale production units of 3000-litre capacity have been used and units up to 10 000-litre capacity are planned. Construction costs are lower for such a tower than for the conventional submerged culture process.

Flavour of vinegar. Whilst the acetic acid in vinegar is responsible for the preservative and solvent properties of vinegar, trace substances contribute to the flavour and improved taste of malt vinegar over a simple solution of acetic acid and water. It is difficult to identify all trace substances, but the major compound is the ester 'ethylacetate' formed from residual alcohol and acetic acid. Studies on vinegar using the gas chromatograph and mass spectrometer have identified 16 esters, 11 alcohols, 11 halogenated compounds, 8 hydrocarbons, 7 carbonyls and 5 ethers or acetals. A further 21 trace compounds were unidentified. The origin of many of these compounds is unknown.

2.4 Other food production processes

Bread. When yeasts have an abundance of O_2 they ferment glucose to CO_2 and water with the production of little or no alcohol (see §2.2). This process is used in the baking industry where yeast is allowed to act on sugar in the dough; the resulting CO_2 forms tiny bubbles in the dough which lightens (or 'leavens') it, giving the bread a more open texture.

Citric acid. This is used in the manufacture of lemonade and is produced by the fermentation of glucose by the mould *Aspergillus niger*. Few commercial details of the process seem available, but it is known that traces of iron are required in the growth medium which has a pH between 1 and 2 at a temperature of about 32°C. Excessive aeration of the medium leads to gluconic acid production in place of citric acid whilst a change in pH to near neutral will favour the formation of oxalic acid!

Glutamic acid. Used in the manufacture of packeted soups (as monosodium glutamate) for flavour enhancement, glutamic acid can be produced by very many microorganisms. However, in recent years the Japanese manufacturers have concentrated commercially on the Gram-positive, pleomorphic, non-sporing bacilli, especially *Corynebacterium glutamicum*, and have found isolates capable of giving high yields of glutamic acid from carbohydrate sources. *C. hydrocarboclastus* will accumulate high yields of glutamic acid from hydrocarbons. Other corynebacteria can produce glutamic acid from acetic acid.

Corynebacteria and other microorganisms can produce a wide variety of amino acids from carbohydrate and other sources. Amino acids so produced include alanine, valine, diaminopimelic acid, lysine, ornithine, citrulline, arginine, phenylalanine, threonine, proline, isoleucine, leucine, tryptophan and aspartic acid.

As in industries using lactic acid bacteria, infection of the producer cultures by phage can have serious commercial consequences. Even mild phage infection can delay fermentation and cause losses of glutamic acid. Additionally, the pH changes and the culture medium foams and becomes difficult to filter, leading to further production losses. Prevention of phage contamination involves scrupulous hygiene in the factory; not only must all the culture vessels and substrates be sterile but, in a process requiring vigorous oxygenation such as glutamic acid production, the air must be filtered to a very high standard of cleanliness. Phage-resistant mutants are sought amongst producer strains and attempts are also made to prevent phage action by reducing the available calcium, since phages need calcium before they can infect and lyse the host bacterium.

2.5 Yeasts as food

Waste brewers' yeast is marketed in most Western countries as a food additive under various brand names as a 'yeast extract'. Yeasts are rich in protein and vitamins of the B group and contain a substantial quantity of fats. During the Second World War a process was devised in Britain for using a food yeast, *Candida utilis*, for protein production, growing it on molasses from the sugar industry. Although this was an acceptable source of protein, with the only real disadvantage that it was a little too rich in vitamin B, the experiments stopped soon after the war partly as a result of consumer resistance to novel sources of food. Even where, in the Third World, there is a desperate shortage of protein for human consumption, two problems have prevented the ready acceptance of yeasts as foods. One is consumer resistance and the other is that the food surpluses from which yeast protein might be derived are thousands of miles from those who would benefit from the protein. In recent years however, world shortages of food have become so acute that the concept of single-cell protein, either for direct human consumption or as fodder for animals destined for human consumption, has been reinvestigated. Studies on the production of single-cell protein, frequently with *C. utilis* (though other organisms are also used), are being made on substrates such as petroleum oils, on by-products such as molasses and on various waste products: for example, as an adjunct to the purification of sewage, *C. utilis* has been successfully grown on a number of industrial effluents including wood hydrolysate, phenolic wastes and peat breakdown products. Whether yeasts from such sources can be regarded as suitable for human consumption is a question currently under debate. A single situation in which water is purified, industrial waste removed and food produced for human consumption, is one which should interest the industrialist and conservationist alike. The whole scope of this subject is too great for a book of this size but it forms one of the growing points in industrial microbiology.

2.6 Production of antibiotics

Antibiotics can be defined as 'chemical substances produced by microorganisms that, in dilute solution, exert a growth-inhibiting effect on other microorganisms'. This definition is a little narrow for it leaves out the sulphonamides which are, and always were, produced chemically but which act in much the same way as antibiotics. The phrase 'in dilute solution' is intended to exclude substances such as alcohol which are toxic in high concentrations but cannot be regarded as antibiotics. Nevertheless the definition is useful.

There are two theories of antibiotic production which ignore their role as aggressive agents towards other organisms. One, which has received most support, is that the selective advantage of secondary metabolism enables the cell to keep enzyme systems in operative order when cell multiplication is no longer possible. It is likely that, in the absence of a substrate, an enzyme system would not be maintained. Thus when one substrate, perhaps even a minor one, is exhausted, a cell might lose all the enzymes of a particular pathway. These would have to be replaced at a later stage when the substrate again became available. Secondary metabolism would keep the system in running order but would have the effect of producing some peculiar compounds.

The alternative theory is that antibiotic production may be a detoxifying process. Thus substances which are formed during metabolism and which might prove dangerous to the cell are modified and excreted. They may remain toxic to other cells however and this would account for the antibiotic action. A corollary of this theory would be that addition of a potentially toxic compound should lead to increased antibiotic production. This is certainly so in the case of penicillin production which, as shown later, can be regarded as the detoxification of carboxylic acids such as phenyl acetic acid by linkage to 6-amino-penicillanic acid (6APA). Only strains which produce penicillin can grow in the presence of significant amounts of phenylacetic acid and the most efficient penicillin producers tolerate the highest amounts of phenylacetic acid.

These two hypotheses are not incompatible, and if both are considered it is easy to see why antibiotic production appears late in the growth cycle (exhaustion of a nutrient) and why the substance has the action that we detect on other microorganisms. Details of the chemical structure of antibiotics and their mechanism of action can be found in HAMMOND and LAMBERT (1978).

2.6.1 Neomycin

Neomycin is a typical antibiotic. It is a product of *Streptomyces fradiae* and was discovered by Selman Waksman in the U.S.A. during a deliberate search for antibiotic compounds. It was the first in a series of compounds now known as the aminoglycosides; the molecules of such substances are formed by the glycosidic linkage of amino sugars to other sugar substances.

Microbiology. *Streptomyces* are aerobic, Gram-positive bacteria which are free-living, filamentous, nonphotosynthetic, rod-shaped organisms, less than 2 µm wide which in young cultures produce a branching mycelium. Chains of conidia are produced from aerial hyphae but sporangia are not formed. Most species have a pronounced earthy odour: the characteristic smell of leaf mould in a wood is the result of the growth of *Streptomyces*.

Production. Prior to 1940 it was known that the *Streptomyces* group produced antibacterial substances, but the full significance of this was not appreciated until after the introduction of sulphonamides in 1935 and penicillin in 1940. Since 1940 over 10 000 cultures have been examined, about 400 antibiotics have been isolated and studied, and about 20 have found their way into commercial use. These include streptomycin, framycetin, the tetracyclines, chloramphenicol, erythromycin and novobiocin as well as neomycin.

Early studies showed that the optimum conditions for growth of *S. fradiae* are a temperature of 28°C, pH 7.0–7.5 and adequate oxygenation. Two distinct growth phases then occur. The first is one of active growth with a high oxygen demand, during which the available carbohydrates are decomposed with a resulting slight drop in pH. The second phase is one of slower metabolic change in which there is a lowered oxygen demand, release of nitrogenous compounds leading to a rise of pH, and lysis of the mycelium. It is during this phase of lysis that neomycin is released into the medium (Fig. 2-3). Neomycin is initially bound to the mycelium, it forms insoluble complexes with ribonucleic acid with maximum complexing at pH 6.2. Above pH 8 there is little or no binding, and this accords with the observations that neomycin is released as the pH rises above the initial value of about 7. Neomycin production (and certainly its release) can be regarded as occurring when exhaustion of fermentable carbohydrate induces the release of nitrogenous material with a simultaneous rise in the pH of the nutrient medium.

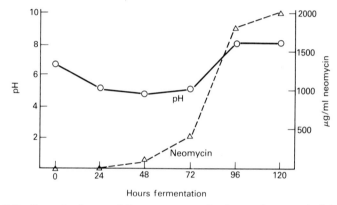

Fig. 2-3 Growth phases of *Streptomyces* and release of neomycin into the medium.

2.6.2 Penicillin

Microbiology. The *Penicillium* species that produce penicillin are moulds – filamentous fungi that are commonly found in our environment. They are frequently found on decaying fruit and vegetables and some are used in the production of cheeses (see §2.1.2). The spores are formed on a flat paint-brush-like structure and may be of many different colours; *P. chrysogenum* (see Fig. 2-4), used in the production of penicillin, has dark green spores.

Production. When Fleming first noticed the action of penicillin in 1929, the *Penicillium* mould was growing on a plate that had been inoculated with colonies of the bacterium *Staphylococcus*. Penicillin formation was therefore known to occur when the fungus was growing on an agar surface with full oxygenation. When commercial penicillin production began at the start of the Second World War all

Fig. 2-4 Colony of *Penicillium chrysogenum*. Note the fluid droplets on the surface that contain the yellow pigment chrysogenin. (This photograph is taken from a slide/tape presentation by courtesy of Camera Talks Ltd.)

therapeutic penicillin was produced in this way, growing the fungus in static culture; broad bottom flasks (Roux bottles), shallow pans and even milk bottles were used. When liquid media were introduced, penicillin was still produced by surface growth. This required thousands of bottles and was an inefficient process, although yields of about 20 units/ml (12 μg penicillin/ml) were obtained. Although the Northern Regional Research Laboratories (NRRL), Illinois, U.S.A., began a search for better strains, none was found to produce more penicillin on surface culture than the original Fleming isolate.

A submerged culture method was devised, however, which was less cumbersome than the surface culture method and one strain, NRRL 832, yielded more penicillin this way than did the Fleming isolate. Yields per ml were low compared with the surface method but the advantages of deep culture outweighed the disadvantages. *Penicillium chrysogenum*, NRRL 1951, isolated from a mouldy cantaloupe melon purchased in a market at Peoria, Illinois, was found to be slightly better than NRRL 832, but the introduction of corn steep liquor (the liquid remaining after maize has been soaked before being ground to flour) raised penicillin production to 100 units/ml. Corn steep liquor acted as a nitrogen source. Studies on variants produced by spontaneous mutation showed NRRL 1951 B25 to be a better penicillin producer than the parent with yields of 250 units/ml (Table 1). Attempts were then made to find even more variants by the use of

Table 1 Selection of mutants for penicillin production.

Strain number	Origin	Units of penicillin per ml of submerged culture
NRRL 832	Natural isolate	Less than 20
NRRL 1951	Mouldy melon	100
NRRL 1951 B25	Natural mutant	250
NRRL 1951 B25 X1612	X-ray mutant	500
WIS 47 1564	Mutation in the laboratory	850
Modern industrial strains	Mutation in the laboratory	100 000

mutagens such as X-rays, ultraviolet (UV) light and nitrogen mustard. One mutant X1612 obtained by X-ray mutation of NRRL 1951 B25 gave 500 units/ml. But all these cultures had one disadvantage. In addition to penicillin, all produced a yellow pigment, chrysogenin, from which the fungus derived its specific name. This colour was very difficult to remove during the purification of penicillin and so all penicillin preparations were contaminated with pigment. Accordingly, even though a mutant was found which produced 900 units

§ 2.6   PRODUCTION OF ANTIBIOTICS   21

penicillin/ml plus pigment, another which produced 850 units/ml of penicillin but no pigment became the standard strain, this mutant was isolated at the University of Wisconsin and was coded WIS 47 1564. Repeated mutation experiments have yielded the modern industrial strains which produce no pigment but up to 100 000 units/ml of penicillin. Submerged culture is still used in tanks of 90 000-litre capacity and all operations are computer controlled.

As with most of these complex substances we do not have a very clear idea how penicillin is formed. According to one theory, the tripeptide theory, amino-adipic acid, cysteic acid and valine combined to form a tripeptide. The tripeptide is partially closed to form ring structures giving iso-penicillin N and this is followed by enzymic

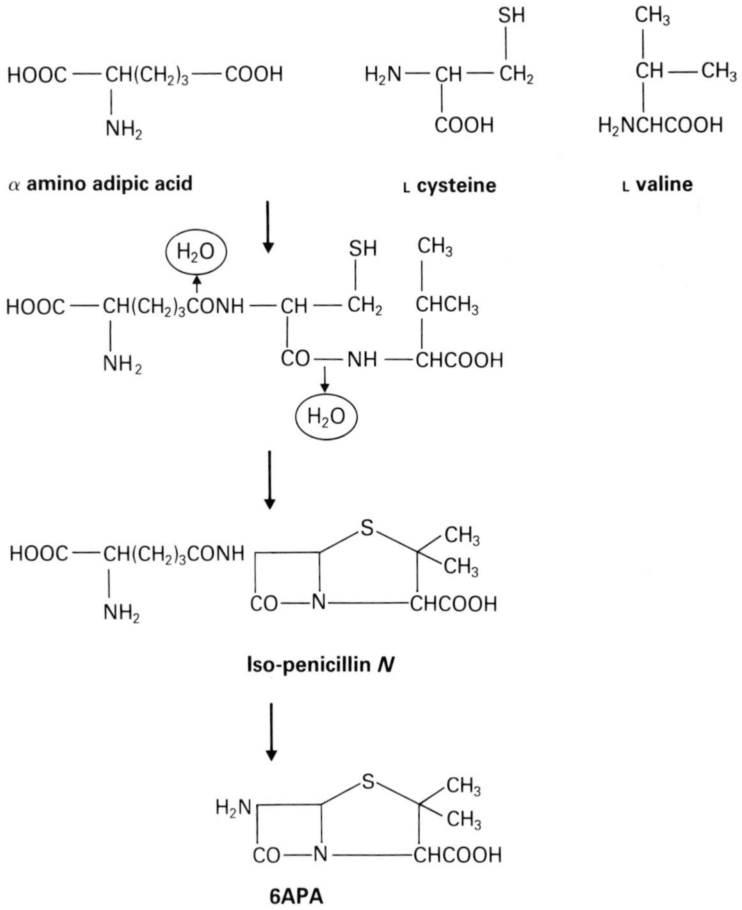

$$HOOC-CH(CH_2)_3-COOH \qquad H_2N-CH(SH)-CH_2-COOH \qquad CH(CH_3)-CH_3$$
$$\qquad\quad NH_2 \qquad\qquad\qquad\qquad\qquad\qquad\qquad\qquad H_2NCHCOOH$$

α amino adipic acid   L cysteine   L valine

$$HOOC-CH(CH_2)_3CONH-CH(SH)-CH_2 \quad CH(CH_3)CH_3$$
$$\qquad NH_2 \qquad\qquad\qquad\quad CO-NH-CHCOOH$$

(with H_2O added and H_2O removed)

Iso-penicillin *N*

6APA

Fig. 2-5 Formation of 6APA according to the tripeptide theory.

conversion to benzyl penicillin or all other penicillins by way of 6-amino penicillanic acid (6APA) (Fig. 2-5).

Semi-synthetic penicillins. In the early stages of large-scale culture four penicillins were formed, known as F, G, K and X. Penicillin G or benzyl penicillin has the most desirable properties and its production can be encouraged by adding phenyl acetic acid to the medium. Benzyl penicillin has the disadvantage that it is unstable in acids and so cannot be administered by mouth, for the hydrochloric acid in the stomach would destroy it. Phenoxymethyl penicillin (penicillin V), developed later, is acid stable and so does not have to be injected. The yields of a particular penicillin can be favoured by putting the appropriate carboxylic acid or side chain precursor in the growth medium. Although many carboxylic acids have been added, and many different penicillins made, all new penicillins are made by starting with 6APA and performing a chemical conversion. Over 2000 new penicillins have now been made of which about ten are used to treat patients.

It is possible to obtain 6APA directly from the fermentation process but this gives a low yield and so is costly; a better method is to use acylase enzymes obtained from bacteria to remove the side chain from penicillin G or V, this yields 6APA about 90% pure. 6APA may also be produced by chemical cleavage of this bond. The addition of various side chains is used to confer special properties: methicillin is resistant to bacterial enzymes that destroy penicillin by opening the lactam ring (known as penicillinase or β lactamase), and is chiefly used against the Gram-positive cocci; ampicillin is resistant to acid and useful against Gram-negative bacilli; cloxacillin is both acid and penicillinase resistant, and so on.

Recent developments in the technology of producing penicillins include the use of immobilized enzymes. Commercial use of enzymes has always been hampered by the difficulty in recovering the enzymes at the end of the batch process. When whole-cell preparations are used there is a danger that other unwanted reactions may occur.

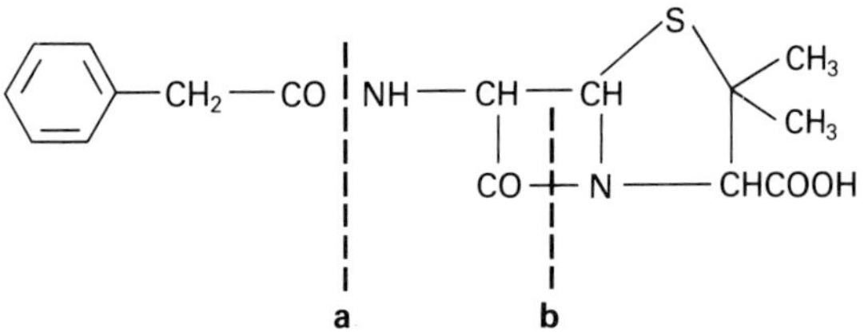

Fig. 2-6 (a) Site of action of penicillin acylase. **(b)** Site of action of penicillinase (β lactamase).

Immobilizing the enzymes by attaching them to some large inert particle such as Bentonite or Sepharose has the advantages that recovery is easy and non-destructive to the enzyme and continuous processes can be used which are more economic than batch processes. The bacterial enzyme penicillin acylase cuts the penicillin G molecule as shown in Fig. 2-5. The same enzyme may serve to hydrolyse the molecule at alkaline pH or to synthesize at acid pH. This is the case with the acylase derived from *Escherichia coli*, though other systems are known in which two separate enzymes are necessary. Antibiotic production has, then, become an area where advances in biochemical engineering are leading the field.

3 Some Results of Microbial Degradation

3.1 Microbial activity in sewage disposal

Where there are only few humans and no industrial activity, it has long been the custom to discharge waste products directly into local rivers and streams. This has been possible because natural, moving waters have the ability, through their aerobic microorganisms, to return organic waste to a condition in which it can again be used by plants – this is known as 'mineralizing' and occurs as long as sufficient oxygen is available. If the biological oxygen demand (BOD) becomes too great however, anaerobic reduction of organic matter takes place with the release of foul-smelling sulphur compounds. It is clear, then, that a sewage works is required to function in the same way as a natural river. The waste must be 'purified', i.e. the organic matter removed, or at least released, in an oxidized form so that subsequent BOD is not high and the water is made fit to return to public use. It is assumed that all pathogenic or disease-causing microbes will be destroyed in the process.

The quality of effluent reaching a sewage works will differ according to the source and the process must be capable of adaptation and change to new substances. Industrial wastes may contain large amounts of potentially toxic substances such as metals or phenols, whilst domestic and farm wastes are principally organic in origin and may have their own particular microbial flora. If surface water from roads is included it may contain tar-based products eluted from the road surface or, in winter, large quantities of sodium chloride (used to reduce icing of the road surface).

3.1.1 The sewage works

Sewage farms. An old method of disposing of sewage was to allow it to run between ridges of soil on which crops were planted. This method permits aerobic breakdown of the organic waste and directly reuses the nutrients made available. However, the capacity of this system is not very great and it has been superceded by other methods.

Modern sewage systems. These use a series of treatments which are outlined in the flow chart in Fig. 3-1. After passing through a grid (filter screen) to remove large particles which might damage pumps, and through a grit chamber to allow grit and silt to settle out, the

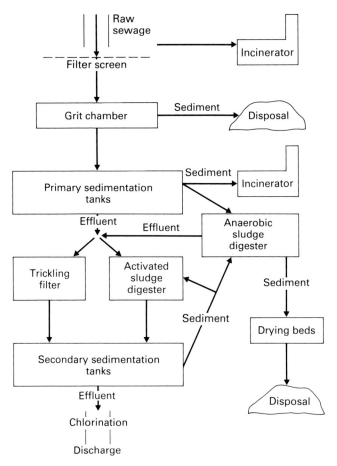

Fig. 3-1 Flow chart of a sewage works.

sewage passes to a primary sedimentation tank. Here some microbial oxidation occurs and the microbial mass plus flocculated organic matter settles out. This material is described as primary or raw sludge. This raw sludge may be dried and burnt, or pumped to anaerobic sludge digesters where the organic material is broken down to organic acids, methane and CO_2. Provided that not too much toxic metal is present the sludge from this stage may be dried and used as a fertilizer. The effluent from the primary sedimentation tank is then trickled through large circular beds of clinker, slag or similar porous material with a large surface area – these are the trickling or percolating filters. As the effluent is sprayed onto these beds and begins to trickle through, further oxidation occurs with considerable microbial proliferation. Microbial populations are kept in check by larvae, metazoa and worms which prey on the bacteria and help to reduce clogging. Any further flocculated material obtained from the effluent is settled out in secondary sedimentation tanks and returned to the anaerobic digestion tanks. The effluent contains salts of nitrogen, phosphorus and sulphur and can safely be discharged into rivers and streams.

As an alternative to trickling filters, the activated sludge process may be used in which material from the primary sedimentation tanks and from the anaerobic digesters is mixed with an inoculum of flocculated material from a previous batch and then oxygenated vigorously by stirring or by the injection of O_2 or air. After oxidation the sediment is removed and treated to further anaerobic digestion after which the effluent is chlorinated and discharged.

Lagoons or oxidation ponds. A process used more frequently in those parts of the world with more constant sunshine than Britain are the lagoons or oxidation ponds. These depend mainly on algae to produce O_2 by photosynthesis. Co-actions between organisms are very complex: at night when the algae release CO_2 in place of O_2, the ponds may become anaerobic; during the day they are highly oxygenated.

Microbiology. The microbiology of sewage is very complex. Some organisms are engaged in breaking down substances and others use the breakdown products to synthesize new compounds. The degree of intimacy required may be judged from the history of the organism formerly known as *Methanobacterium omelianskii*. Although thought to be pure cultures, the original isolates appeared pleomorphic; initially they were able to form methane from ethanol, but during anaerobic cultivation in an atmosphere of hydrogen and CO_2 the ability to utilize ethanol was rapidly lost. When incubated in an ethanol H_2/CO_2 mixture a slender rod-form organism was obtained which produced methane from H_2 but could not use the ethanol (strain MoH). When incubated in ethanol N_2/CO_2 a shorter wider rod was isolated able to produce small quantities of H_2 but not methane (S strain). When the two were mixed together, ethanol was converted to methane. Other evidence confirms that these two organisms grow symbiotically on alcohol, the S organism fermenting the alcohol to H_2 and acetate and the strain MoH using the H_2 to reduce CO_2 to methane; the removal of H_2 is important because it is toxic to the S organism. Microbial co-actions are therefore of great complexity and, in describing briefly the role of organisms in sewage disposal, it is possible to present only a greatly simplified picture.

With the exception of the anaerobic bacteria which are solely responsible for methane production, it is difficult to assign a specific role to a particular group of organisms for many different microorganisms possess enzymes capable to degrading specific substances.

3.1.2 Bacterial breakdown of waste materials

Lists of the various microorganisms found in sewage and summaries of enzyme production are not meaningful unless a specific type of waste is considered in a particular sewage plant under known conditions of temperature and rate of flow. The following accounts suggest how two particular products, cellulose and detergents (surface active agents) may be broken down and lists some of the microorganisms which may be involved in the breakdown processes. Details of enzyme pathways, must be sought elsewhere since we shall be concerned only with overall reactions.

Breakdown of cellulose. Cellulose forms a major part of all plant material; for example, cotton fibres are about 90% pure cellulose and straw is about 35%. Cellulose forms approximately 4% of all sewage or about 14% of the settled fraction of raw sludge.

A single cellulose polymer molecule consists of over 3000 $\beta(1,4)$-linked glucose residues (Fig. 3-2). The ability to degrade this, and other similar polysaccharides, is widespread amongst microorganisms and occurs, for example, in the genera *Pseudomonas*, *Clostridium* and *Bacteroides*, which are all major components of the sewage microflora. Many of the fungi found in trickling filters can also break down cellulose.

Enzymes involved in cellulose breakdown are known as cellulases; they are many and complex, and are most frequently attached to the cell wall. Cellulases first of all degrade cellulose to carboxy-methylcellulose, or 'reactive' cellulose, then further degradation results in the formation of the dimer cellobiose which is finally degraded to glucose.

Fermentation or oxidation of glucose may then be carried out intracellularly by a variety of microorganisms to give simple organic compounds such as lactic, acetic or formic acids or ethanol (note that in Fig. 3-2 two different notations are given for glucose). Short-chain compounds are then degraded or reduced, for example by *Methanobacterium* species, to H_2, CO_2 and CH_4 (methane).

Breakdown of surface active agents. Surface active agents (surfactants) are composed of molecules which have both water- and lipid-soluble portions. They therefore tend to aggregate at oil/water interfaces, reduce surface tension and permit emulsification. In

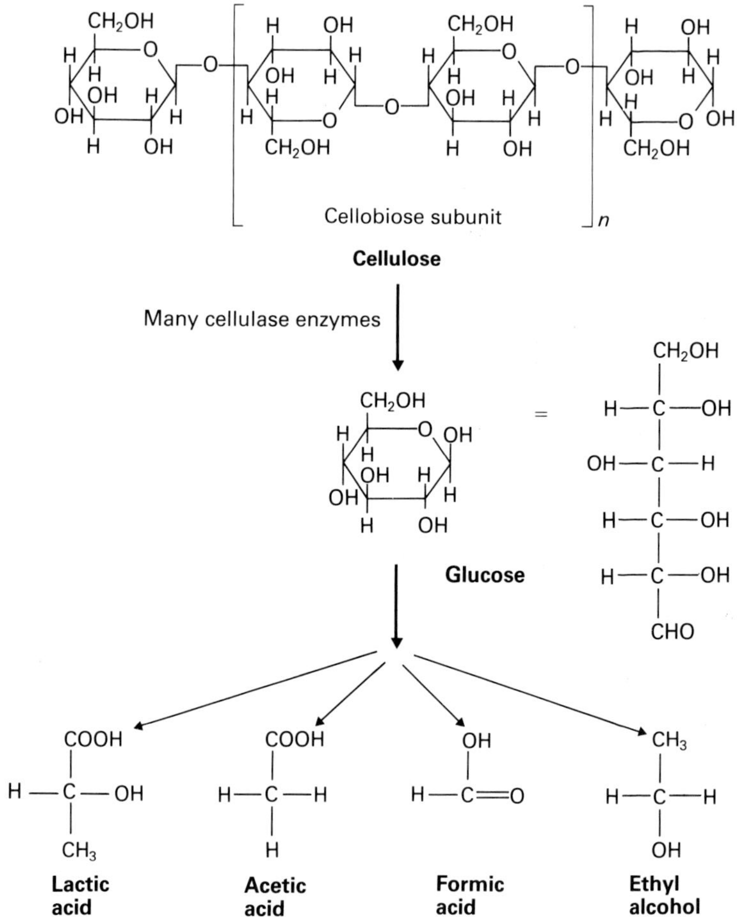

Fig. 3-2 Degradation of cellulose.

consequence they are used as cleansing agents and the two most familiar types are soaps and detergents. Soap surfactants have been in use for many thousands of years, they have the general formula RCOONa, where R is an odd number of carbon atoms, usually 15 or 17. Amongst detergents 'secondary alkyl sulphates' have the general formula $ROSO_3H$ of which the industrial detergent Teepol® is an example and 'alkyl benzene sulphonates' the most widely used detergents have the general formula

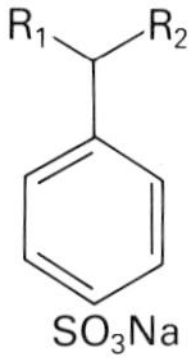

$$\underset{SO_3Na}{\underset{|}{\overset{R_1\diagdown\;\diagup R_2}{\overset{|}{C_6H_4}}}}$$

where R_1 and R_2 are short chains usually totalling about 12 carbon atoms. Today these are generally straight chains which are more easily biodegradable than the branched chains formerly used.

Pathways involved in the degradation of surfactants are many and varied, but in general desulphonation occurs outside the bacterial cell to yield sodium sulphate and the corresponding alcohol. This may then be further degraded intracellularly. Soaps may precipitate as the insoluble calcium and magnesium salts, but are fairly easily biodegraded by β oxidation:

$$CH_3CH_2CH_2CH_2COOH + O_2 \xrightarrow{\beta \text{ oxidation}} CH_3CH_2COOH + CH_3COOH$$

These simple fatty acids are then degraded as suggested earlier to CH_4, CO_2, H_2O etc. Many microorganisms possess sulphatase enzymes able to degrade organic sulphate to inorganic sulphate plus the corresponding alcohol:

$$RCH_2OSO_3Na \rightarrow Na_2SO_4 + RCH_2OH$$

The alcohol is metabolized to the organic acid and this is further degraded by β oxidation.

Alkyl benzene sulphonates are degraded in a similar way by desulphonation and β oxidation of the side chain with terminal oxidation of the benzene ring.

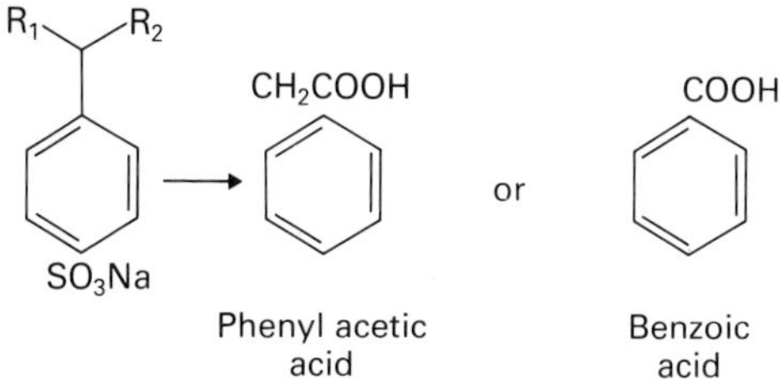

$$\underset{SO_3Na}{\underset{|}{\overset{R_1\diagdown\;\diagup R_2}{C_6H_4}}} \rightarrow \underset{\text{Phenyl acetic acid}}{C_6H_5\text{-}CH_2COOH} \quad \text{or} \quad \underset{\text{Benzoic acid}}{C_6H_5\text{-}COOH}$$

The benzene ring may be degraded by a variety of routes involving further oxidation to form the diphenols and subsequent cleavage of the ring adjacent to one hydroxyl group. One example is shown here.

$$\underset{\text{CH}_2\text{COOH}}{\bigcirc} \rightarrow \underset{\text{HO}}{\bigcirc}\underset{\text{OH}}{\overset{\text{CH}_2\text{COOH}}{}} \rightarrow \text{COOH-CH=CH-CO-CH}_2\text{COOH}$$

with products:

$$\begin{array}{c} \text{COOH} \\ | \\ \text{CH}_2 \\ | \\ \text{CH}_2 \\ | \\ \text{COOH} \end{array} \qquad \text{CH}_3\text{COCH}_2\text{COOH}$$

3.1.3 Role of fungi in sewage disposal

Fungi can be as efficient as bacteria in the removal of organic compounds from sewage, but in general they are undesirable as dominant members of the flora. Fungal film is tougher and breaks up less easily than a corresponding film of bacteria and so may cause clogging of filters, resulting in the production of anaerobic conditions. The pH of the sewage may be very important in controlling the amounts and species of fungus present; the prevalence of, for example, *Geotrichum candidum* may be related to occasional flushes with sewage of pH 5.0 or less. Industrial wastes are especially associated with a predominance of fungi, *G. candidum* has been shown to grow abundantly in a sewage plant which received waste of pH 3.0 from an acid plating works. In sludge treatment plants, the predominant species are those found in raw sludge, though *Trichosporum cutaneum* may become particularly abundant on drying beds.

Predaceous fungi are the most interesting members of sewage plant flora and *Dactylella bembicoides*, a nematode-trapping species, has been isolated from a nematode-rich activated sludge. Other predatory species recovered from sewage are *Arthrobotrys oligospora*, *Harposporium auguillulae* and species of *Trichothecium*.

3.1.4 Role of algae in sewage disposal

Although algae may be seen covering the upper layers of percolating filters, they play only a minor part in the purification process and may even reduce the efficiency of the filter. It has been calculated that algal photosynthesis can only account for about 5% of the O_2 needed by other microorganisms in the filter.

In contrast, algae play a most important part in the purification processes in oxidation ponds or lagoons. The most common genera are *Chlorella, Scenedesmus, Chlamydamonas* and *Euglena*. Small ponds with a retention time of about one week depend on algae for O_2 produced during photosynthesis. In return, the bacteria provide CO_2 and organic or inorganic nutrients for the algae; pH can again be seen to have a profound effect, for example in a sewage plant receiving milk waste from a dairy plant, the bacterial species *Streptococcus lactis* and *Lactobacillus* were responsible for the production of mildly acid conditions (pH 5.8) following lactose fermentation; as a result the alga *Selenastrum* was selected. When the acidity was neutralized the algal flora changed to *Pandorina, Navicula* and *Oocystis* which are adapted to milk waste.

3.2 Wash-day enzymes

Specific enzymes that have been used very extensively in the home are those put into 'biological detergents' or 'enzyme detergents'. In 1971 over 70% of all detergent products contained these enzymes but use has fallen steadily since that time. The enzymes were used to remove protein stains, for example milk or blood, from clothing and, because they were used in laundry, had some special properties. Most laundry is carried out at alkaline pH and high temperature and the enzymes used were optimal at pH 10 and 60°C. They were derived from the Gram-positive, spore-bearing bacterium *Bacillus subtilis* and were proteases known as subtilisins. The term protease is given to enzymes able to hydrolyse peptide bonds.

$$\underset{\text{Protease}}{NH_2CHCO-NHCHCOOH} + H_2O \rightarrow NH_2CHCOOH + NH_2CHCOOH$$

(with R groups on each CH)

One protease used commercially in enzyme detergents has a molecular weight of 27 600, contains 274 amino acid residues and is a single peptide chain; in addition to acting on proteins this enzyme also has esterase activity. It might have seemed impossible to find an enzyme which would fit so well into the range of temperature and pH used in laundry yet here is one seemingly designed with wash-day in mind. At 37°C the protease has a pH optimum of 10 but exhibits 50% or more activity over a pH range of 7 to 11.5. At a constant pH of 8.5 the temperature optimum is between 60°C and 65°C but 50% or more activity occurs over the range 48°C to 70°C. Doubtless many other enzymes exist which would find a household use if only they could be produced economically, but the many stages of growth, removal of microorganisms and purification of the enzyme make this a facet of microbiology for the future.

3.3 Biodeterioration of materials

In this section it will be possible only to give a general idea of the complexity of the role of microbes in deterioration. This is partly because of the enormous number of organisms involved in, for example, the rotting of timber, but also because in many cases the chemistry of the processes has not been worked out, and we can only guess at the mechanisms from a knowledge of the end products.

Aircraft fuel systems. Prior to the introduction of jet-engined aircraft a recurrent fuel problem in tropical countries was attack of sulphates by *Desulphovibrio* species leading to the formation of sulphides which dissolved in the fuel. The major problem in jet fuel is the growth of a filamentous fungus *Cladosporium resinae* in the fuel tanks. This can lead to blockage of the filters by a 'black sludge' composed of dirt and rust particles bound together by fungal mycelium. Fuel tanks are located in the wings of jet aircraft and so are subject to heating to ambient temperatures up to 80°C when the plane is standing on the ground under a tropical sun. Temperatures of $-40°C$ are common in flight at altitudes of 9000 m or more. Cold does not kill the fungus, it merely slows the growth, whilst spores of *C. resinae* have been shown to survive 80°C for one hour.

Water needed for growth is rapidly precipitated from jet fuel as the temperature decreases during flight, but is not redissolved so quickly when the fuel warms again. Additional water is derived from condensation when warm air enters the cold tank as the aircraft descends. Water and jet fuel do not mix so that water droplets are formed in the fuel and these tend to cling to the coating inside the tanks. As microbial growth occurs the lining of the tank itself may be attacked releasing nitrogen and sulphur compounds. More than half of the volume of jet fuel consists of paraffins of which the short-chain (C_8 to C_{14}) components are utilized by the fungus to provide energy as well as a carbon source.

Biodeterioration of metals. Any type of microorganism may be responsible for certain types of corrosion in metals by forming a 'concentration cell'. When a metal is immersed in water, metal ions pass into solution leaving a net negative charge (electrons) on the metal. These electrons can react with water and oxygen to form hydroxyl ions which, if the metal is iron, give rise to rust. This reaction ceases when an equilibrium is set up between the metal and water which limits the escape of the metal ions. If however the equilibrium is destroyed, the reaction proceeds indefinitely. The presence of microbial colonies on the metal surface can prevent the equilibrium being reached, for example by sulphate-reducing bacteria which remove hydrogen to form hydrogen sulphide. Thus corrosion proceeds faster in the presence of microbial colonization even if the microorganisms have no special ability to attack the metal.

Some bacteria cause biodeterioration of metals by a more direct method. Reference has already been made to *Cladosporium resinae* in aircraft fuel tanks. During growth of this mould short-chain hydrocarbons are attacked and yield organic acids, which may react with the metal or lining of the tank.

A third form of biodeterioration, and one that is important in metals buried in soil, for example gas pipes, is due to sulphide or sulphuric acid production. *Thiobacillus*, which are aerobic, straight Gram-negative rods can oxidize sulphides to free sulphuric acid and can live and metabolize at a pH of 0.7. The effect of sulphuric acid at this concentration on metals does not need to be emphasized. *Desulphovibrio*, which are anaerobic, free-living, curved, rod-shaped,

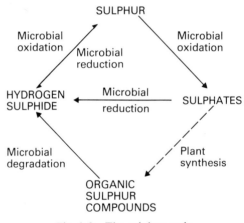

Fig. 3-3 The sulphur cycle.

Gram-negative cells with a single polar flagellum, are able to reduce soil sulphates to sulphides or to hydrogen sulphide. The formation of iron sulphide is detrimental to the strength of the iron pipe. Corrosion of this type can be retarded by adequate insulation of the pipe or by providing a 'backfill' in the pipe trench of sand or chalk to ensure good drainage. Even the coatings and wrappings of buried pipes may be subject to attack by *Pseudomonas, Streptomyces* or *Aspergillus* species. *Desulphovibrio* are also responsible for the foul smell of mud at the bottom of ponds with a high B.O.D., due to the release of hydrogen sulphide.

Biodeterioration of timber. An example of wood rot is the deterioration of beech and Scots pine used in the construction of power station cooling towers. Basidiomycetes such as *Poria, Peniophora* and *Lentinus* cause white rot and brown rot which result in structural failure of the uprights and cross members. White rot results in wood which may disintegrate to a lint-like consistency due to attack on the lignin and cellulose. Brown rot attacks the cellulose and carbohydrate leaving the lignin untouched. It can be recognized by the dark colour and the characteristic way that the wood breaks up into small brick shaped pieces along and across the grain.

Eighteen other fungi including ascomycetes, such as *Chaetomium globosum*, and moniliales, such as *Monodictys putredinis*, have been recorded from timber in cooling towers. Too little work has been done on this problem to form a coherent picture; in nature there may be great differences between apparently similar situations. In tests on Scots pine in sea-water, twelve fungus species were recovered in Devon but only six from Hampshire. The reasons for this appear to relate to differences in water temperature and degree of pollution.

Bacteria often attack wet timber. *Bacillus* species (Gram-positive, spore-bearing rods), especially *B. polymyxa*, are able to hydrolyse pectin and hemicellulose which are abundant in the middle lamellae of plant cell walls. Wood borers amongst the crustacea such as *Limnoria, Sphaeroma* and *Chelura* have been shown to possess cellulose-digesting bacteria in their gut and it is assumed that these bacteria contribute to the nutrition of the crustacea.

Biodeterioration of cellulose (see also §3.1.2). It is, of course, not surprising that microorganisms should exist which are able to break down cellulose since a high proportion of plant tissue is cellulose and this would not otherwise be recycled. Experiments with cotton fabrics buried in soil show that the fabrics lose their strength within a week and may be completely destroyed within a month. A more domestic example of cellulolytic microorganisms are the mould fungi which attack cellulose in the insoles and stitching of shoes with the result that the shoes fall apart.

Biodeterioration of wool. Superficial growths of bacteria such as *Bacillus subtilis* or *B. cereus*, or of fungi such as *Aspergillus* or the Actinomycetes, are collectively known as mildew and may discolour or degrade wool long before there are changes in the strength of the fibre. Such attack is generally started on impurities such as vegetable oils or soaps for these provide a readier source of nutrient. However, the protein of wool itself may be attacked and broken down finally to amino acids. Although only about 10% of the wool fibre can be attacked in this way it is sufficient to downgrade its quality for commercial purposes.

3.4 Biodeterioration of pollutants

Degradation of pollutants is a fortuitous form of sewage disposal. Man is very careless with his products or by-products of manufacture and the environment would be even more polluted than it is, were it not for the action of microorganisms.

It is now known that the disappearance of pesticides from the soil is due to microbial degradation. The chemistry of the various processes has yet to be fully worked out, but it is known that the combined action of two (or more) bacteria is sometimes needed: for example, the insecticide DDT appears to be broken down by the joint action of *Aerobacter* and *Hydrogenomonas* species. Organophosphates (also insecticides) are broken down by *Thiobacillus thiooxidans*.

Hydrocarbons are broken down by a number of organisms including the fungi *Botrytis cinerea* and *Penicillium glaucum*, the yeast *Candida utilis* and, amongst the bacteria, *Pseudomonas, Nocardia* and *Mycobacterium*. This should not surprise us, for long-chain hydrocarbons exist naturally in plant material and also in the cuticle of insects such as the cockroach. In small amounts, hydrocarbons are degraded and the elements recycled without fuss. When we talk of oil pollution however, we think in terms of large amounts of oil being spilled in an accident such as damage to an oil tanker or to an oil rig. There seems little doubt that even these spillages could be dealt with by microorganisms given time, for crude oil spillages on fertile land in temperate climates are more or less completely degraded in two to six months. However, in cold climates, or in deserts where the soil flora is poorly developed, the pollution may last for years. About five million tonnes of oil are spilt at sea each year and may in a very short time contaminate beaches and kill seabirds and other marine life. Current microbial degradation of oil spills is too slow to cope with major spillages, but perhaps some way of speeding this process can be found just as the yields of antibiotics and other productive processes have been increased.

3.5 Microbial deterioration of food

Food spoilage can be divided into three convenient categories. The first is the obvious spoilage caused by a visible growth of microbes, usually moulds or pigmented bacteria, for example on bread. It is curious that we will delight in 'blue' cheese which is covered in mould (*Penicillium roqueforti*) yet refuse to eat a piece of bread which is 'contaminated' with very similar moulds. The second type of spoilage is that which is less immediately obvious but becomes evident by slight discolouration, smell or altered texture, for example sour milk or 'ropy' bread. Finally, it is possible for food to be heavily contaminated with, for example, *Salmonella* sp. or *Staphylococcus aureus* yet show no external signs of contamination. Only when the food is eaten and food poisoning results do we realize that microbial growth or toxin production has taken place.

It is the second group which is microbiologically the most interesting, for food spoilage is only another example of the way in which microbial enzymes act on substrates. The main degradative processes caused by the growth of microorganisms in food are set out below.

Carbohydrate foods	+ carbohydrate fermenting organisms	→	acids + alcohols + gases
Protein foods	+ proteolytic microorganisms	→	amino acids + amines + ammonia + hydrogen sulphide
Fatty foods	+ lipolytic microorganisms	→	fatty acids + glycerol

In fresh or preserved food the particular spoilage observed will depend upon a number of factors: the kind of contaminating organisms, chemical composition of the food, availability of nutrients, temperature of storage, oxygen supply, moisture content, osmotic concentration, pH, presence of inhibitory substances. The range of microorganisms involved in food spoilage is correspondingly diverse.

3.5.1 Spoilage of fruit and vegetables

Mechanical damage by bruising or wounding or infection by pathogens may leave fruit or vegetables more susceptible to subsequent attack by spoilage organisms. Once spoilage microorganisms have established growth in the product, spread occurs by contact with intact fruit and vegetables causing increased wastage.

The type of spoilage is dependent on the product and on the attacking organism. If the food is soft and juicy, the rot will tend to be soft and mushy and some leakage of the fruit or vegetable contents may occur. In some instances the mycelium of the infecting mould may have penetrated the interior of the fruit or vegetable but have scarcely damaged the epidermis. In other cases the mycelial growth covers the outside of the food and may be brightly coloured.

Each fruit or vegetable has a characteristic type of decomposition and consequently particular kinds of microorganisms will predominate. The names are usually descriptive of the appearance or occasionally of the contaminating organism.

Bacterial soft rot. The causative agents are *Erwinia caratovora* and related species as well as *Pseudomonas* species. Almost all vegetables which lack a high acid content are susceptible to these bacteria, but carrots, celery and potatoes are the most commonly affected. The organisms secrete enzymes which break down pectin in the plant cell wall causing cell disruption and breakdown of the tissue in the area infected. This gives rise to the characteristic softening associated with these bacterial infections. The tissue breakdown facilitates further spread of the organism and also may provide a focus for invasion by other organisms.

Mould rots. As the name suggests these rots are due to fungal moulds. Often the names of the rots themselves, for example Alternaria rot, blue mould rot, identify the attacking organism directly by name or by colour of the spores.

The majority of injured fruit and vegetables develop grey mould rot caused by the fungus *Botrytis* which is recognizable by its characteristically grey coloured mycelium. *Penicillium* sp. (blue mould rot) cause about 30% of all fruit rot.

Citrus fruits tend to be prone to attack by *Alternaria*, and *Rhizopus* causes soft rot of fruit other than the citrus species. Vegetables such as tomatoes, potatoes, carrots and cabbages are susceptible to downy and powdery mildews due to species of *Peronospora* and *Phytophthora*. *Fusarium*, *Diplodia* and *Phomopis* are collectively called stem end rots and are responsible for the deterioration of bananas, sweet potatoes and citrus fruits by, as the name suggests, attacking the stem ends. Figs, dates and grapes seem to be most susceptible to attack by *Aspergillus* and in particular *A. niger* which produces dark brown spores. The spore masses give the mycelium a black colour, hence the name 'black mould rot'.

As already mentioned, mechanical damage or injury to the fruits and vegetables leaves the way open for invasion by microorganisms. To prevent spoilage and wastage, care needs to be taken during harvesting, packing, transporting and selling. Soil-borne microorganisms may be present at the time of storage. Contaminating organisms can be removed from the food by washing but this is unsatisfactory as washing can also spread microorganisms and any remaining moisture may encourage microbial growth. Chlorine or dehydro-acetic acid is added to the water as both are effective in decontaminating food surfaces and preventing microbial spread. Sodium orthophenylphenate is also an effective decontaminating agent when used as a dip. Other chemicals known to inhibit microbial growth or destroy spoilage organisms are impregnated in wrapping papers or sprayed onto the fruit and vegetables. The method of storage is another important factor in preventing spoilage. Temperatures need to be low enough to slow down microbial growth, but freezing must be avoided unless the fruit and vegetable are to be prepared as 'frozen food' since the formation of ice crystals causes mechanical damage and can result in autolysis. The use of 'inert' gases, for example N_2 and CO_2, in the storage atmosphere also cuts down rotting by inhibiting the growth of microorganisms as well as slowing down other biochemical changes that normally occur in the product.

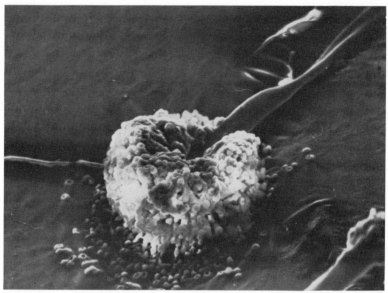

Fig. 3-4 Scanning electron microscope picture of *Aspergillus niger* spore head. The mop-like appearance is characteristic and is the feature from which the genus is named. The mycelium and some spores have collapsed as a result of drying prior to examination in the vacuum chamber of the microscope.

3.5.2 Spoilage of cereals and bread

Grain crops should not spoil or deteriorate as a result of microbial attack if they are prepared and stored correctly, in a *dry* atmosphere, since moisture is the key factor in spoilage. The presence of a little moisture will permit growth of moulds such as *Penicillium* and *Aspergillus* and a lot of moisture will encourage the growth of coliform and lactic acid bacteria and yeasts. Bacteria ferment the grain products to acid and, at a suitably acidic pH, yeasts start growing and complete the fermentation to alcohol.

When cereals are baked as bread any vegetative microorganisms present in the dough are killed by the high temperatures. However, spores may survive baking and further contamination may occur during slicing, wrapping or handling of the bread. If the loaves are wrapped too soon after baking, without having sufficient time to cool properly, condensation will occur in the wrapping and this encourages microbial growth. Storage in warm, damp conditions also aggravates deterioration.

In general there are two main types of microbial spoilage of baked bread – mouldiness and ropiness. The chief moulds involved in spoilage are frequently called 'bread moulds', although they include *Penicillium* and *Aspergillus* sp. (Fig. 3-4) found on other foods.

A number of methods are employed to prevent mouldiness of bread. They include the removal of spores from the air by filtration or UV-irradiation of the bakery rooms; condensation promotes microbial growth, so prompt and adequate cooling of the loaves is necessary before wrapping. If the bread is to be sliced prior to wrapping, the slicing equipment needs to be sterile to prevent recontamination. Formerly vinegar and acetate were added to the dough to inhibit fungal growth, but now sodium or calcium propionate (mycostatic chemicals) are usually employed.

Ropiness in bread is recognized by an unpleasant odour together with discolouration and a slimy, sticky consistency that allows the bread to be stretched into strings when broken. The condition is rare in commercially baked bread because of the aseptic measures used. However, ropiness often occurs in bread made at home, particularly during hot weather.

The organisms responsible for ropiness are variants of *Bacillus subtilis* or *B. licheniformis* which are present during baking. They are encapsulated bacteria, and production of a polysaccharide capsule causes the sticky condition. The bacilli also produce extracellular enzymes which hydrolyse the bread proteins and convert starch to sugars, thus contributing to rope formation.

In order to prevent ropiness it is necessary to minimize, or eliminate, contamination of the dough or baked bread by *Bacillus* spores. The preventative measures employed to reduce contamination are similar to those for the prevention of moulds. *Bacillus* spores require hot humid conditions for germination and this is presumably why ropiness most often occurs during the summer months. Germination and growth of *B. subtilis* spores is inhibited by an acid pH. The use of a dough formula that will result in a pH of about 5 provides a good means of stopping any growth if spores do happen to be present. The acid pH of the bread is accomplished by addition of acetic, tartaric, citric or lactic acid to the dough ingredients.

3.5.3 Spoilage of sugar-containing products

Very few products with a high sugar content are spoiled by microbial attack due to their high osmotic concentration. Any bacteria present are plasmolysed and consequently killed. Spoilage of sugar compounds is therefore limited to that caused by osmophilic microorganisms such as species of *Leuconostoc, Bacillus*, yeasts and certain moulds able to withstand and grow at high osmotic values.

Raw cane or beet before processing is not high in sugar and consequently can be attacked by a number of spoilage organisms. *Leuconostoc mesenteroides, L. dextranicum* and species of *Bacillus* form gum and slime in the raw cane and during sugar manufacture. Various moulds (e.g. *Aspergillus, Cladosporium*) may also be responsible for decomposition and gum formation in the sugar. The gum may cause fouling of the pipes, pumps and valves in a refinery.

Once manufactured the sugar is quite resistant to microbial attack, but, if damp, can still be spoiled by moulds such as *Aspergillus* and *Penicillium*. Some yeasts and moulds are able to degrade the sucrose to glucose and eventually to acids or CO_2 and water.

Preserves, such as peaches or pears which are canned in syrup, are subjected to heat during the canning process so only osmophilic organisms that are able to withstand the heat treatment will cause spoilage. When the can is opened and the syrup is left exposed to the air for a time, mould eventually grows over the surface. Similarly, surface mould growth may occur in bottled or canned syrup if air is left in the bottle or can and the contents contain contaminating spores.

3.5.4 Spoilage of fermented products

These include olives, various pickles and sauerkraut. Sauerkraut depends for its flavour on the biochemical processes of a variety of microorganisms. If these organisms are inhibited in any way, for example by too high or too low a temperature, any undesirable bacteria present will be encouraged to grow and so alter the flavour. If the fermentation process is carried on too long, growth of gas-forming *Lactobacillus brevis* is promoted together with *L. plantarum*. Both produce 'off-flavours' and the latter, due to encapsulation, produces slime and causes ropiness. Fermented pickles, for example cucumbers, may also be spoiled in a variety of ways. Floating or bloated cucumbers ('floaters' or 'bloaters') are caused by the presence of gas inside the cucumbers produced by species of yeast or by *L. brevis*. The growth of encapsulated bacteria in the product causes 'slippery' pickles, and 'soft' pickles are made so by bacteria which produce the enzyme polygalacturonase responsible for the degradation of plant pectin. If the contaminating bacteria are hydrogen sulphide producers, combination of the gas with iron in the pickling water results in a black deposition of ferrous sulfide on the pickles, forming 'black' pickles. The colour may also be due to black pigmented bacteria such as *Bacillus nigrificans*.

3.5.5 Spoilage of canned foods

During the canning process, products are subjected to very high heat treatment so that if any spoilage does finally occur in the can, it is probably due to heat-resistant spores of thermophilic bacteria. In particular, some putrefactive anaerobic bacteria of the genus *Clostridium* have spores that are highly resistant to heat. On germination and growth the bacteria cause putrefaction which is often accompanied by swelling of the can due to the production of CO_2 and H_2 gases. One of these organisms, *C. botulinum* is the cause of botulism – a severe form of food poisoning which is often fatal. The spores of *C. botulinum* however, are unable to germinate in an environmental pH lower than 5. The hazard is therefore minimized by the canning of high acid foods, i.e. foods with a pH of 4.7 or lower. Further, foods in the acid range generally require a considerably lower heat processing to effect preservation, the processing schedule being designed to kill all vegetative cells. Canned foods of low acid content (e.g. cereals and legumes) are particularly prone to attack by *Bacillus* spp. which produce organic acids in the product, thus giving the food a sour taste though no gas is produced.

Defects during the canning process, such as leakage, may contribute to food spoilage by recontamination. Microleakage does occur on occasions permitting the water used for cooling to enter the can. Thus the use of clean, preferably chlorinated cooling water is a highly desirable quality control procedure. An imperfect closure can also allow the entry of bacteria which may be pathogenic. Though extremely rare, such occurrences have been reported, for example the recontamination of canned tuna with *C. botulinum*. The Aberdeen typhoid fever outbreak (1964) occurred when people ate imported canned corned beef that had been contaminated with *Salmonella typhi* after processing. The cans apparently had been cooled in polluted river water. Staphylococcus food poisoning has occurred as a result of recontamination of a can of green peas during cooling.

3.5.6 Spoilage of milk products

Milk. Milk is an ideal nutritional medium for many kinds of microorganisms. It has a neutral pH and is rich in microbial foods. Carbohydrate is present in the form of lactose, butterfat and citrate; nitrogenous compounds include proteins, amino acids, ammonia and urea. Although sterile at its source, milk usually contains a number of different species of microorganisms. *Micrococcus caseolyticus, M. freudenreichii* and *Streptococcus liquefaciens* are frequent inhabitants of freshly drawn milk. But the influence of sanitary conditions (e.g. cleanliness of milking equipment and the cow's coat), airborne contaminants, and the actual handling of the milk, all contribute to the presence of other bacteria in the milk, such as *Bacillus* sp. and *Escherichia*.

Efficient cooling of the milk on the farm will prevent spoilage before it is pasteurized. When held at room temperature the normal milk flora of *Streptococcus lactis* together with coliform bacilli, lactobacilli and micrococci (the homofermentative organisms) will produce lactic acid together with small amounts of acetic acid, CO_2 etc. At higher temperatures (over 39°C), *S. thermophilus* and *S. faecalis* will commence fermentation and will be followed by *Lactobacillus bulgaricus* (heterofermentative) to make the milk very acid. The milk then coagulates to give the jelly-like curd which is the basis of cheese production.

Even after pasteurization, spoilage due to souring and acid formation may still occur because of the growth of lactic streptococci. Pasteurization eliminates the majority of acid-formers, but the heat-resistant spores of *Clostridium* and *Bacillus* species are able to survive and are therefore the main causes of gas formation which is evident by foam at the top of the milk or bubbles caught in the curd. Curdling accompanied by the development of foul odours is characteristic of milk stored at refrigerator temperatures for long periods. This is due to the presence of psychrophilic bacteria (e.g. *Pseudomonas*) which are able to multiply successfully at temperatures below 4°C.

Butter. The major spoilage in this case is rancidity. This is due to the formation of free fatty acids from the butterfat glycerides by lipase enzymes of *Pseudomonas, Proteus, Achromobacter, Bacillus* and other organisms.

Cheese. Contaminating coliform bacteria such as *Escherichia* or *Aerobacter* present a problem in the production of cheese. These bacteria ferment the lactose in the cheese liberating CO_2 which produces gas holes in the cake. Lactobacilli and streptococci also produce angular open areas in the cheese by formation of CO_2.

In the home, however, deterioration is most often caused by the surface moulding; many different species of mould may grow on the cheese. Moulds are usually aerobic, so the exclusion of air in the packaging aids in the prevention of mould growth on the cheese surface.

3.5.7 Spoilage of meat and fish

Meat is high in moisture, rich in nitrogenous foods, well supplied with minerals and growth factors and, since there is usually some form of fermentable carbohydrate present, is a complete medium for the growth of a diverse range of microorganisms.

Aerobic spoilage. Species of *Pseudomonas, Achromobacter, Streptococcus, Bacillus* and *Micrococcus* are all responsible for surface slime. It has been estimated that about $3-5 \times 10^7$ bacteria per 1 cm^2 of meat surface are present when slime and smell first becomes apparent. Yeasts also produce slime as well as causing lipolysis resulting in off-odours and tastes.

Discolouration of meat is another form of spoilage and is due either to metabolic compounds produced by the microorganism or to pigmentation of the microorganism itself. *Lactobacillus*, for example, produces oxidizing compounds and these change the meat colour from red to shades of green or brown. But usually surface colour changes are due to pigmented bacteria: *Serratia marcescens* causes 'red spot'; *Pseudomonas syncyanea* gives the surface of the meat a blue colour; *Micrococcus* or *Flavobacterium* turns the meat yellow; *Chromobacterium lividum* gives greenish-blue to brownish-black spots; *Photobacterium* produces phosphorescence due to its luminosity. Many of these are short rod form bacteria (see Fig. 3-5).

Bad-smelling or ill-tasting meat is often the result of bacterial growth on the meat surface and is usually evident before any visible signs of spoilage. The acid products of bacterial fermentation (e.g. formic, acetic, butyric and propionic acids) are responsible for the sour taste and odour in bad meat.

Moulds are another problem in aerobic spoilage. Their growth on the meat surface causes stickiness, or fuzziness may occur on the meat

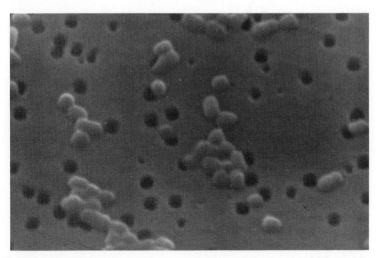

Fig. 3-5 Scanning electron microscope picture of short rods which are about 1 micrometre long by 0.5 micrometre wide. The structure of the filter, 0.45 micrometre holes in a smooth sheet of material, is clearly seen.

exterior due to mycelial growth. Pigmented moulds may also cause discolouration, for example *Cladosporium* causes black spot. Surface spoilage of meat by moulds however is not very extensive and in most cases the mould can be trimmed off without any harm to the rest of the meat. But extensive bacterial growth over the surface is often accompanied by deeper penetration and diffusion of microbial compounds.

Anaerobic spoilage. The chief types of anaerobic spoilage are 'souring' and putrefaction and the organisms involved are the facultative and anaerobic bacteria. 'Souring' applies to the offtaste brought about by the presence of formic, acetic, butyric, propionic and other fatty acids in the meat. All are produced by bacterial action. 'Putrefaction' is most commonly caused by *Clostridium* species and refers to the anaerobic decomposition of protein, accompanied by the production of foul-smelling compounds such as hydrogen sulphide, mercaptans, indole, ammonia etc.

Fish autolyses more easily than does red meat and organisms from the fish gut invade the viscera soon after death. The microbiology of fish is basically the same as that of water so that the Gram-negative bacilli *Pseudomonas*, *Achromobacter* and *Flavobacterium* predominate.

Fish have a high content of non-protein nitrogen and autolysis makes available nitrogenous materials such as amino acids and amines as well as sugars; these all support microbial growth. Bacteria form trimethylamine (which causes the typical fishy odour), ammonia, amines such as putrescine and cadaverine, fatty acids and aldehydes. Eventually sulphides and mercaptans, most of which have offensive odours, are formed. Discolouration of fish may be caused by marine pseudomonads (greeny-yellow) or *Serratia* (red) or by *Micrococcus* species which have a deep yellow pigment.

4 Diseases of Man and Animals

4.1 Food poisoning and other gastrointestinal diseases

Bacterial diseases transmitted by faecal contamination include typhoid and paratyphoid fevers, dysentery, cholera, and *Salmonella* infections. The agents which cause these infections do their damage locally, whereas others, particularly the parasitic worms, spread from the intestinal tissues to other parts of the body. All leave the host in excreted faecal matter and enter the next host via the mouth. There are also several virus diseases, the agents of which may enter the body through the mouth and produce gastrointestinal symptoms. Examples of these are hepatitis, epidemic gastroenteritis, and epidemic diarrhoea. The control of all these diseases depends on good sanitation and personal hygiene.

4.1.1 Salmonella food poisoning

The story of Typhoid Mary – an American cook who was a faecal carrier of *Salmonella typhi* and infected many families before she was eventually isolated and removed from food preparation – is well known but worth summarizing here as an illustration of the danger from even a single carrier.

Typhoid Mary, whose real name was Mary Mallan, first attracted attention in 1906 when General William Henry Warren with his family and servants, a total of 11 people, rented a seaside house in Oyster Bay for the summer. Mary Mallan was engaged as a cook and in less than three weeks there were six cases of typhoid in the house, but no others in the whole of Oyster Bay. George Soper, a sanitary engineer, was called in to investigate. He was eventually able to show that Mary Mallan had been responsible for at least 14 other cases of typhoid whilst employed as a cook during the previous five years. Mary was arrested and held at a hospital where it was repeatedly shown that she excreted *Salmonella typhi* in her faeces; she was a carrier of the organism but was herself completely healthy. She was eventually released after giving an undertaking not to work as a cook, but eight years later George Soper was called to investigate an outbreak in the Sloan Hospital for Women in New York where he found that Mary had returned to her old trade and had infected more than 20 people. Mary Mallan spent the rest of her life in a house in the grounds of a hospital, for she could not be cleared of the typhoid bacilli and she was too dangerous to be allowed to return to normal life. Soper could trace 53 cases of infection, three of whom died, directly to Mary Mallan and it is believed that an outbreak which reached 1500 cases was started by her. People like Mary Mallan, who become faecal carriers of *Salmonella* species, are a danger to the community for they do not feel ill and are usually unaware that they are a source of infection. That Mary should choose to be a cook was a tragedy, for it gave her the maximum opportunity to transmit the disease to others. Although it might seem that this old story would serve as a sufficient warning on this danger to health, British newspapers of July and August 1976 carried stories of a cook with paratyphoid who had infected passengers on a cruise ship in the Mediterranean.

Another outbreak of *Salmonella* food poisoning (though not typhoid) illustrates other facets of the problem. In September 1973, in mid-western America, 125 of 173 people who had eaten a meal in a certain restaurant developed diarrhoea, severe abdominal pains and other symptoms, on average 23 hours after the meal. Eleven of the people had to be taken to hospital for treatment. Faecal cultures from 18 of those affected yielded three *Salmonella* species. Faecal cultures from five of the eight restaurant employees also grew salmonellae but none were recovered from the remaining food. However, a study of the food eaten by the victims suggested that potato salad and chicken dressing were the vehicles of transmission. These had been prepared in pans which had previously contained uncooked chicken pieces which may have been contaminated with *Salmonella*. Feed samples from the farms from which the chickens originated contained *S. typhimurium* and *S. cubana*, implying poor production hygiene and suggesting that other *Salmonella* species might have been present in other batches of feed. It was necessary to close the restaurant for decontamination. The economic cost of the outbreak was estimated at about £15 000.

In October 1977, newspapers carried reports of a massive outbreak of *Salmonella* food poisoning in Stockholm: more than 1000 school children and teachers became ill after eating ham salad served in school cafeterias. The food was prepared in a central kitchen and distributed to several schools in western Stockholm. The main fear of local health officials was that parents and other family members would catch the infection and so cause a gigantic epidemic.

The way to control infection of this type is to break the chain of transmission. It can be broken at many places; for example, feed stuffs should not contain *Salmonella*, this would reduce the potential load of infection. However, the best place to interrupt transmission is at the food preparation stage. Food which is to be eaten raw or cooked but cold, must not be allowed to come into contact with other potentially contaminated food; kitchen utensils must always be kept clean and thoroughly washed after every use. Staff handling food must maintain high standards of personal hygiene.

§ 4.1   FOOD POISONING AND OTHER GASTROINTESTINAL DISEASES   47

In 1975 food poisoning cases in England and Wales increased by nearly 40% to reach a total of almost 12 000 reported cases. Published figures are not available for later years but data from the Communicable Disease Surveillance Centre shows that there was a drop of nearly 1000 cases in 1976, a further reduction in 1977 seems possible. About 74% of cases over the last ten years have been due to *Salmonella* organisms, many of the infections may be acquired as a result of eating in restaurants but infection can also be acquired from eating at home. An increase in salmonellae in poultry feed has led to increased contamination of chickens and ducks.

4.1.2   Staphylococcal food poisoning

This description of food poisoning can be completed by describing the 'Jumbo Jet' outbreak in 1975. This was due to staphylococcal food poisoning which does not depend on the presence of live organisms in the food, as does *Salmonella* food poisoning, but on an extracellular toxin which is heat stable. Anyone who ingests food containing the toxin may become ill, even if the food has been recooked, and illness may occur between 30 minutes and about 6 hours after consuming the food, depending on the dose of toxin absorbed and the susceptibility of the person.

On February 3rd, 1975, a Boeing 747 (Jumbo Jet) left Tokyo for Paris with intermediate stops at Anchorage (Alaska) and Copenhagen. At Anchorage the plane took on a new crew, plus breakfast and snacks for the flight to Copenhagen. Both meals had been prepared in Anchorage two days before the flight and kept in storage. By the time the plane reached Copenhagen many passengers were ill and the flight was held up for a long time. Figure 4-1 shows the number of passengers becoming ill at intervals after breakfast. Passengers were suffering from diarrhoea, vomiting, nausea and many had high temperatures. In all, 194 of the 344 passengers and 1 steward from a crew of 20 were ill and at Copenhagen 142 passengers and the steward were admitted to hospital for short periods.

Studies of this outbreak showed that the toxin was carried in ham which had been made up into an omelette and reheated on the plane. The ham had been contaminated by a cook in Anchorage and had not been stored at a low enough temperature. The staphylococci grew and produced toxin before the meals were taken onto the plane.

This outbreak had some far reaching consequences. It was usual for crew members to eat the same food as the passengers. Fortunately, in this incident, none of the *flight* crew was ill, but the problems of trying to land a Jumbo Jet whilst suffering from diarrhoea and vomiting can be imagined. The World Health Organisation convened a meeting which discussed the composition of menus and the training of

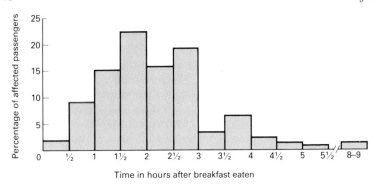

Fig. 4-1 Histogram showing times at which 192 passengers fell ill after eating breakfast. The times for 2 passengers are unknown.

personnel in food hygiene, but especially important was the firm recommendation that flight crews should receive meals from two different sources so that the chance of all being affected during one flight is reduced to insignificant proportions.

4.2 Aerosol infections

Transmission by aerosol infection is responsible for the spread of a wide variety of bacterial and viral diseases. The pathogenic organisms are carried from person to person in microscopic droplets of airborne saliva or on dust particles. Diphtheria, tuberculosis, streptococcal and pneumococcal infections are bacterial diseases transmitted in this fashion. Amongst the viral diseases so carried are smallpox, chickenpox, measles, poliomyelitis, epidemic influenza and the common cold. The control of infection transmitted by exhalation of droplets is difficult. The only sure way to prevent such spread would be to require that all individuals wear masks equipped with filters. But such an extreme measure has rarely been found practicable or enforceable.

When a person coughs or sneezes a fine spray of saliva is expelled which breaks up into individual tiny droplets. The larger droplets fall quickly to the floor and are removed from circulation but the smaller ones evaporate to form the so-called 'droplet nuclei'. These may be up to about 20 µm in diameter but many are very much smaller. Any bacteria, including *Mycobacterium tuberculosis*, in the saliva and sprayed out in the droplets, will be contained in these droplet nuclei. Small particles settle only very slowly even in still air; for example particles of about 14 µm diameter take 1 minute to settle 30 cm through still air. Most occupied rooms have air currents which are capable of keeping such particles airborne for long periods and particles of 14 μm can travel extensively in occupied buildings. The inhalation process must be considered too. The nose is a very efficient sieve and removes most particles from inhaled air. We breathe about 10 000 litres of air a day and most of the particles in that air are removed in the nose and swallowed, to be destroyed in the acid of the stomach. However, very small particles, those less than 5 μm, may pass through the nose and into the lung, those about 1 μm in diameter may be impacted onto the surface of the lung and, although most are usually swept out by the mucous sheet, some remain.

4.2.1 Tuberculosis in man

Tuberculosis is a bacterial disease of man and some animals. The disease is endemic and still a major cause of death in many parts of the world. The agent responsible is the tubercle bacillus (*Mycobacterium tuberculosis*) and the disease is characterized by a process of destruction and replacement of normal tissues with tubercles which may produce both local and constitutional (e.g. high temperature) reactions. The lungs are most commonly infected, but any tissue of the body may be affected, usually as a result of secondary spread. In man the disease symptoms generally consist of pleurisy and vague chest pains, often with coughing, fever, fatigue and loss of weight. The patient may become anaemic and nervous and the sputum be bloodstreaked or frank haemorrhage may occur.

The tubercle, a small nodule produced by the body cells in response to the presence of *M. tuberculosis* and its products, is the basic lesion of tuberculosis. The tubercle may grow in size producing large abscesses which drain pus, spreading the bacilli, or it may become walled off and calcified in healed lesions. Infecting bacilli multiply unobstructed when first introduced and may be caught temporarily in the lymph nodes; this leads to enlargement of the nodes which often undergo the caseous necrosis typical of tuberculosis. The bacilli can then remain alive for many years inside the lymph nodes which gradually become calcified. Alternatively, the bacilli may travel through the lymph tissue eventually reaching the thoracic ducts and then the general circulation; most organisms do not find locations suitable for development and are destroyed, but some remain in microscopic foci and may give rise many years later to infections of bones, joints, lungs, skin and other organs.

We must distinguish between simple infection by *M. tuberculosis* and the appearance of clinical signs of disease. Many people who are exposed to infection do not develop signs of disease which cause them to consult a doctor. Nevertheless they develop resistance to further attack and are then said to be 'immune'.

Microbiology. Mycobacteria are slender, straight or curved rods, measuring about 3 μm by 0.3 μm. They usually occur singly or in clusters but may occasionally exhibit branching and filamentous forms, hence their name. The bacilli are difficult to stain with the usual microbiological dyes and accordingly differ from other bacteria in that they cannot be considered simply as Gram-negative or Gram-positive. However, they stain readily by the Ziehl-Nielsen technique (initial staining with hot basic fuchsin followed by washing with acid and alcohol). In contrast to other bacterial cells, about one third of the tubercle bacillus cell is lipid (fat). Because of this they are not decolourized by the acid-alcohol and are therefore termed *acid-fast* organisms.

These bacteria contain large amounts of protein materials as well as lipids and a variety of polysaccharides to which some of their antigenic specificity may be attributed. They have above-average resistance to ordinary disinfectants, mineral acids and alkaline agents, and are also resistant to drying, being known to survive for long periods outside the human body. However, the organisms are destroyed by exposure to sunlight, and by heat sterilization, including pasteurization.

Transmission. The sputum of persons with tuberculosis is the most common source of organisms. Consequently the disease is usually acquired by inhaling tubercle bacilli expelled by coughing or sneezing from persons with active disease. Of course the risks of airborne infection are considerably increased by the ability of the organism to survive for months outside the host, for example in dust and in books.

Individuals who have stainable tubercle bacilli in their saliva have been found to transmit the bacilli to the air more readily than individuals who have tubercle bacilli in purulent sputum, but not in the saliva. Subjects with tuberculous infections of the larynx, and consequently tubercle bacilli in their saliva, are the most dangerous because they cough incessantly. It will be obvious that, since *M. tuberculosis* most commonly starts its infective process in the lung, small particles, probably containing only one bacillus, are the most infective for man.

Small epidemics usually arise from infections caused by a single individual. An epidemic amongst school children in Denmark was caused by a school teacher who produced tubercle bacilli in her sputum and saliva only when she had a virus common cold. In this instance more than 70 tuberculin-negative students were infected. A school epidemic in Whitesboro School District in New York State was traced to one bus driver. This epidemic was more severe and among 228 children who rode the bus for less than 40 days, 108 became infected. Further, out of the 108 infected, 42 developed active pulmonary infection.

Control. Success in the treatment of tuberculosis depends upon measures which support the patient's own acquired resistance to the disease. Treatment consists of bed-rest, good ventilation and a nourishing diet. Sometimes surgery is required to remove or collapse an infected lung. Most valuable, however, is the use of chemotherapeutic drugs (antibiotics) which have saved or prolonged many lives. Due to the chronic nature of the disease and the walling off of the foci of infection, by the development of fibrous tissue, treatment must be extended over a year or more.

The most effective drug now available is isoniazid (isonicotinic acid hydrazide), introduced in 1952 and now commonly referred to as INH. Other drugs used are streptomycin (SM) and p-aminosalicylic acid (PAS); both introduced in the 1940s. PAS increases the effectiveness of other medications. Generally two or three of these drugs are given together in order to reduce the chance of the mycobacteria developing resistance during the long period of treatment.

The vaccine used for immunization against tuberculosis is made from attenuated live cultures of a bovine strain of *M. tuberculosis*. This strain, which has a low and relatively fixed degree of virulence, was isolated in 1908 by Calmette and Guerin of the Pasteur Institute, hence the name Bacille Calmette-Guerin and the designation BCG for the vaccine. Attenuation – reduction in the virulence for a particular host – is maintained by growing the cells on media containing bile. The vaccine was first used on children in 1921. Since 1951 large-scale immunization programmes have been carried out and more than 250 million BCG vaccinations have been given to children.

Tuberculosis could eventually be eliminated if every active case were diagnosed, isolated and treated. In Western countries there is a declining incidence of, and death from, the disease, due to the isolation of infected persons and treatment with the new chemotherapeutic drugs. The routine tuberculin testing of school children has greatly contributed to the detection of isolated carriers of tubercle bacilli in the community. About six weeks after a primary infection with *M. tuberculosis* the body's defence mechanisms become extremely sensitive to the bacillus proteins. If a small amount of these proteins is now injected into the body, an 'allergic' reaction is elicited involving swelling and discolouration. This forms the basis of the 'tuberculin test' and has great value in locating early infections when treatment is most effective. The tuberculin test is carried out with a heat-killed concentrated filtrate of *M. tuberculosis* (old tuberculin) or with purified protein derivative of tubercle bacilli (PPD). The Mantoux test is the most commonly used form of tuberculin test. It is performed by injecting 0.1 ml of a 1:100 dilution of old tuberculin into the skin of the forearm. The test is then read 48–72 hours after the intradermal injection. A skin reaction of 8–10 mm or more in diameter represents an allergic response to tubercle bacilli and in young people is taken as indication of an infection. In adults and, of course, in some children, it is simply an indication of a previous tubercle infection which is now healed.

4.2.2 Foot and mouth disease of cattle

An airborne virus disease of animals is foot and mouth disease. It rarely causes disease in man but it is a source of economic loss to farmers because infected cattle lose weight, milk production is reduced and, in Britain, the cattle are usually slaughtered. Vaccines do not give protection for more than about four months so that every cow, pig, sheep and goat would need to be vaccinated three times a year. This would be very expensive and it has so far proved cheaper to stamp out the small outbreaks by slaughtering the diseased cattle and any others close enough to have been infected. The disease is difficult to control because the virus may be windborne over large distances.

Foot and mouth disease virus contains RNA and belongs to the picorna group. Since there are very many sub-types the choice of a vaccine is very difficult. Virus lesions in the mouth and on the tongue appear as vesicles (blister-like raised spots) which burst releasing millions of virus particles into the saliva. Between 10 and 100 infectious doses are excreted every minute by infected cows and as many as 10 000 infectious doses are excreted per minute by pigs infected with certain strains of the virus. Aerosols form from the saliva and the airborne particles contain virus. Since many of these airborne particles are very small, less than 3 µm in diameter, they can remain airborne for long periods and virus has been traced 2.5 km from its source, even on a dry day with only light variable winds. Overland, the maximum distance for virus dispersal is between 120 and 150 km, but over the sea, where there are no obstacles, the virus may be carried up to 1000 km. Thus cattle on the east coast of England may be infected with virus which originates in France, Belgium or the Netherlands.

Infection of animals can occur as a result of inhalation of virus particles or as a result of ingesting virus on grass, but this requires 10^4 times as many virus particles as inhalation. Most virus particles are retained in the upper respiratory tract. After inhalation there is an incubation period of about 1 to 5 days and then lesions appear in the mouth, there is an increase in body temperature, a loss of appetite and greatly increased salivation; the cycle of infection is ready to start again.

Human tuberculosis and bovine foot and mouth disease are both airborne diseases but each retain some special features. In human tuberculosis only a very small number of infective particles of the appropriate size are produced, and only one is needed to set up infection. Immunity following infection or vaccination is life-long. Control can be achieved by fairly simple isolation procedures. In foot and mouth disease enormous numbers of infective particles are produced and these infect herds at a great distance from the origin. Immunity following vaccination is of short duration. Thus, even when there is the same portal of entry – the upper respiratory tract – there may be very great differences in the behaviour of different pathogens and the response of different hosts.

4.3 Contact infections

There is a small number of pathogens for which the portal of entry is the skin or mucous membranes. The group includes the causative agents of the venereal diseases, syphilis and gonorrhoea which are transmitted by sexual intercourse or from an infected mother to her baby at child-birth. Three other bacterial diseases transmissible by direct contact are anthrax, tularemia and brucellosis. These are transmitted to man from animals. Virus diseases spread by direct contact include warts and verrucas and herpes simplex (*Herpesvirus hominis*) most familiar as 'cold sores'. Control in all cases is the simple avoidance of direct contact and contact with contaminated material such as floors or clothing.

4.4 Vector-borne diseases

Lastly, an interesting adaptation by some pathogens is the ability to exist in two or more alternative hosts. The plague bacillus for example can multiply in rats, fleas and man; it is thus guaranteed a continuous supply of fresh hosts. Yellow fever, a viral disease, is transmitted to man by a mosquito, with monkeys acting as an alternative host, and rabies is transmissible by a variety of carnivorous mammals. The malarial parasite is also transmitted by a mosquito. Control in this case is accomplished by the elimination of the vector (the species which transmits the pathogen) or the reservoir of infection (the species from which the vector derives the pathogen). Man has thus been able to control and sometimes eradicate such diseases from large areas in this way: bubonic plague has been controlled over most of Europe, yellow fever is no longer endemic in the Caribbean islands and malaria has virtually disappeared from the Mediterranean area.

4.4.1 Rabies – a viral disease of man and animals

Introduction.   Rabies is a fatal, neurotropic virus infection mainly of dogs and other carnivorous animals. Human rabies is caused by the bite of a rabid animal, and it is characterized by a variable incubation period, hyperexcitability, paralysis and, inevitably, death. The disease has a world-wide distribution. In temperate zones it can occur at any season of the year, but is more prevalent during the spring and summer, probably because domestic animals are allowed greater liberty then and cover more territory.

Age, sex and race do not appear to be significant in contraction of the infection and about 50% of the people bitten by rabid animals develop rabies if they are not treated. The circumstance of being bitten by a rabid animal, plus the location of the bite and its character determines the probability of clinical infection. Bites about the face and neck are the most dangerous and more often result in infection, probably because the incubation period of the disease varies with the site of the bite. On average, the incubation period is 30 days for bites on the head, 40 days on the arm and 60 days on the leg.

The clinical manifestations of the disease in man can be divided into three stages. The first, the premonitory period, is characterized by headache, nausea, mental depression and a temperature of 38–39.5°C lasting two to four days. Further symptoms consist of numbness and tingling about the site of the wound, sensitivity to lights and noise and increased salivation, perspiration and lacrymation. During the next stage of the disease – the excitement stage, the patient becomes nervous and irritable and the sight of food or liquids or the sound of running water cause severe spasms of the muscles of swallowing. These symptoms gave rise to the name 'hydrophobia' (morbid fear of water). Most patients die during this phase, but if they survive the convulsions and progress into the paralytic stage, the muscle spasms cease and the patient becomes unconcious; death is then generally caused by respiratory paralysis or heart failure.

History. The first recorded reference of rabies in animals was by Aristotle *c.* 335 B.C. In Britain the disease in dogs was recognized as early as A.D. 1026 and in France in A.D. 900. In the American continent, rabies was recorded in 1750 and has remained endemic since.

Pasteur, in 1881, undertook the first set piece of virus research and showed the virus to be present in the brains of rabid animals. Four years later Pasteur was the first to use preventive vaccination against the disease. In 1904, Adelchi Negri discovered granular inclusion bodies in the cytoplasm of the brain cells of rabies infected animals. These 'bodies' now bear his name and are always associated with the virus-infected cells.

The last case of human rabies contracted in Britain was in 1902; however, since 1945 there have been 13 reported cases in persons contracting the infection overseas. Of these cases, two were in 1975, one in 1976 and two in 1977 and four of these five were from the Indian subcontinent.

Infectious agent. The agent responsible for rabies is a helical, enveloped, RNA virus. The particle is characteristically bullet-shaped, but rod-shaped virions may occur, it is quite large, 100–150 nm in length, and is not readily filterable.

Electron microscopy of the rabies virus reveals delicate but prominent surface projections on the virion. These bear the antigenic determinants of the virus. The transverse cross-striations represent the nucleocapsid or viral core. Essentially the nucleocapsid consists of ribonucleic acid and protein in close association and arranged in a helix (Fig. 4-2). All these features are characteristic of the rhabdovirus

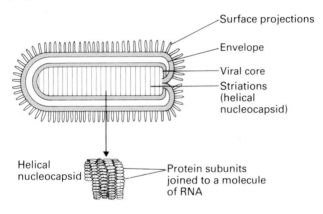

Fig. 4-2 Diagrammatic version of structural protein of virus of rabies group.

viral group of which rabies is a leading member. Other members include vesicular stomatitis virus and Marburg virus.

On infection the virus is believed to travel from its point of entry along the nerves or perineural lymphatics to the brain. After multiplying in the brain, where it does extensive damage, the virus travels to the salivary glands and other parts of the body, probably again along the nerves. Pasteur referred to the virus present in nerve tissues of animals developing the disease under natural conditions, as 'street' virus, whereas the same virus after many serial passages through the brain tissue of rabbits was called 'fixed' virus. Street virus has a long incubation period (15–30 days) and is highly infective. 'Fixed' virus has a lower infectivity rate and a shortened incubation period (6–8 days). Infection with 'fixed' virus gives rise only to paralytic rabies and Negri bodies are not formed in the infected cells.

However, infection with 'street' virus is clinically characterized by either highly excitable or paralytic rabies and Negri bodies are almost always found.

Transmission. Most cases of rabies result from the bite or licking of a wound by a rabid animal. Besides the domestic dog and cat, foxes, wolves, mongooses, skunks, weasels, stoats, squirrels, racoons, opossums and bats are all potential transmitters of the rabies virus to man. Rats and mice do not appear to be important vectors. Rabies is endemic in foxes of the Arctic zone and Germany. In West Germany alone, 2071 foxes were among the 2660 wild animals found positive for rabies in 1963. Wolves and jackals are important vectors in Eastern Europe where many cases of the disease have occurred in man. The role of canine carriers of rabies is fully discussed in KAPLAN (1977).

In South America today, it is the vampire bat *Desmodus rotundus marinus* that is responsible for the severe paralytic disease in man as well as in domestic livestock. The bats, although they themselves show no signs of infection, carry the virus in their saliva and prey on animals and man, especially children. These bats are interesting in that they are the only animals that transmit rabies as true symptomless carriers. Bats tend to fight amongst themselves and so the transmission chain remains unbroken. Cave-dwelling bats have also been known to spread the virus by aerosols or airborne infection and many cave workers have died from rabies with no signs of being bitten, or having an open wound at the time the illness was contracted.

Birds may also be affected but are only very rarely responsible for transmitting rabies to man. The transmission of infection from one of the usual reservoirs to a non-biting host, such as man, results in death of the host and a break in the chain of infection. But in the natural biting host, excretion of virus in the saliva of rabid animals and the high diversity of susceptible animals, ensures the maintenance of the disease in the species. The dog is the animal in closest contact with man and so provides the usual source of human infection. Human to human infection is extremely rare but it can occur. Human saliva contains the virus and salivation is usually excessive in human cases. Hospital staff are therefore at risk and even minor cuts or abrasions of the skin are a danger, so contact with the patient's saliva must be avoided.

Rabies may be the only virus disease in which uniform lethality seems necessary for the survival of the virus in nature. This is simply because only a very extensive infection of brain cells can provide the necessary gross change in temperament required for transmission of the virus. Indiscriminate aggressiveness against other animals is usually quite abnormal in any free-living mammal, but it is only such behaviour that will allow the sequence of infections to be maintained. The long incubation period of rabies in the dog (three to seven weeks) and the presence of virus in the saliva from three to four days before the onset of symptoms, provide ample time after infection for the carnivore to move away from the region where the infective bite was received and find another potential transmitter.

Control. Eradication of rabies can be achieved only when the total elimination of the reservoir of animal infection is possible. It has been maintained in Britain so far by rigid quarantine laws.

No person having developed clinically recognizable rabies has ever been known to recover and, despite prompt and apparently adequate treatment shortly after exposure, clinical rabies has still occasionally developed. Both active and passive immunization are administered following a bite. Rabies antiserum is given immediately after the bite and 24 hours later a course of vaccine injections is started. Semple's vaccine is generally used. It is prepared from the brain tissue of infected rabbits in which the virus is completely inactivated by phenol at 37°C. The Fermi-type vaccine may be used as an alternative; this is prepared from the brain tissue of infected sheep in which the virus is only partially inactivated with phenol at 22°C. In this case a small amount of active rabies virus remains. Both vaccines however, sometimes give rise to severe neuroparalytic reactions.

General immunization against rabies is not yet recommended except for bitten persons. For people such as veterinarians, dog handlers, field naturalists and laboratory workers, two types of vaccine are now available almost devoid of neural side reactions – the duck-embryo vaccine and the high egg passage. Flury chicken-embryo vaccine in dogs three months of age or older, produces excellent immunity which lasts at least three years.

Immunization, together with the registration and licensing of dogs, is one of the most important weapons in rabies control. There should also be rigid enforcement of quarantine regulations and control of the domestic dog population with the elimination of stray dogs and the confinement for seven days or more of any dog that has bitten a person.

According to W.H.O., outbreaks of rabies develop most often when the species of animal that transmits the disease has become unusually abundant. Steps should therefore be taken to reduce the number of such animal populations. Persons should be warned not to handle sick animals or pets that behave strangely, and the control of the vampire bat is also advised. Finally, public education, especially of children, is recommended emphasizing the need to avoid and report all animals that appear sick, and legal authorities must be encouraged to prosecute persons seeking to abuse the quarantine laws.

5 Diseases of Plants

5.1 Mechanisms of plant disease

Fungi and viruses are the microorganisms most frequently involved in plant disease; although bacteria cause a few serious diseases, the overall economic losses are relatively small. Virus diseases may cause a general loss of vigour in a plant, perhaps by interfering with photosynthesis. Fungus diseases may result in blocking of phloem or xylem tubes, either by fungal hyphae or by outgrowths of the plant tissue itself; such blockage leads to wilting of the plant as the water supply is cut off. Sometimes toxin formation occurs which results in death of the whole plant. Many plant diseases are multifactorial.

The enzyme systems described earlier in this book in relation to the synthesis or degradation of specific compounds are frequently also active in plant disease. An example of this may be seen in the pectinase enzymes involved in wilt or soft rot of plants.

Pectins are complex polysaccharides formed from anhydrogalacturonic acid units in a chain-like structure. Plant pathogens produce a number of pectin-splitting enzymes such as pectinesterase, which hydrolyses methyl ester groups in pectinic acids; polygalacturonase which hydrolyses the 1.4 linkages between adjacent sugar molecules; and other enzymes collectively known as depolymerases. Great care must be taken in attempting to apply data obtained in one host/parasite system to another, for each enzyme system has a different optimum pH (indeed it is probable that these different optima help to restrict parasites to one host or group of plants). The pectinase of *Sclerotium rolfsii* has a pH optimum between 3 and 5.5 whilst the pectinase from *Botrytis cinerea* has an optimum at pH 6.2. Pectins occur in plant cell walls. The first recognizable membrane between cells, known as the middle lamella, appears to be composed mainly, perhaps entirely, of pectins. The next layers, the primary and secondary cell walls, have much pectin although the main wall substance is cellulose. The chief effect of pectin-splitting enzymes on plant tissue is, therefore, to separate cells along the line of the middle lamellae. This is followed by death of the cells. Although this generalization can be made, there are many poorly understood differences between the various pathogens and their action on plants.

Toxin production contributes to pathogenicity in plants; for example, the browning and death of leaves in Dutch elm disease may be the result of toxin production in the woody tissue. However, it is more difficult to be certain of toxin production in plants than in animals or man. In animals there is usually an antigen/antibody response which can itself be used to study the course of an infection. Plants do not form antibodies in response to infection and this has made a study of toxins in infection more difficult in plant pathology. It is necessary to make clear the distinction between antibodies, produced by animals in response to infection, and phytoalexins which are produced by plants. Antibody production is a host response which serves to protect the animal against further infection with the same pathogenic organism, the response is specific to the pathogen. Phytoalexins are chemical substances produced by plants in response to infection, but they are non-specific and serve rather to protect the plant against infection with other pathogenic microorganisms.

5.2 Transmission of plant pathogens

5.2.1 Wind transmission of fungal parasites

The fungi usually produce spores in vast quantities. Spores range between 3 and 30 µm in diameter and because they have so large a surface for their mass, they fall very slowly in still air and are easily shifted by small currents of air. Understandably then, aerial dispersal is one of the most important ways in which spores are disseminated. Wind dispersal is rapid and spores of the rust fungus *Puccinia graminis* for example, have been found at 4240 m over infected grain fields. The wind carriage of three well known fungal parasites is given in Table 2.

Table 2

Parasite and host	Rate of fall	Distance carried by an average wind velocity of 32 km/h
Alternaria (affects apples, pears)	3 mm/sec	4800 km
Puccinia graminis (wheat rust)	12 mm/sec	1200 km
Helminthosporium (corn blight)	20 mm/sec	704 km

The spread of a fungus disease in a crop is controlled by many factors. Consider, for example, a foliage disease of an annual crop caused by a spore-forming fungus. A spore settles on a healthy leaf and germinates and then the mycelium begins to grow through the leaf tissue. After a while it begins to form spores which are dispersed; some fall on new leaves and the cycle begins again. It is the time scale of these events that is important. If the plant variety is very susceptible to the pathogen, the process of mycelium formation and sporulation will be more rapid than in a resistant or partially resistant variety. If the pathogen is a virulent one too, then the growth process may be fast. Weather conditions tending to favour the pathogen, the closeness of planting in rows, the availability of other fields of susceptible plants – will all contribute to the production of an epidemic of infection rather than a single isolated incident.

Details of the way in which the fungus mycelium actually penetrates the leaf surface can be found in DEVERALL (1969) and WHEELER (1976). The end result of a successful infection in a plant or group of plants is the release of vast numbers of spores. A mature wheat field may carry $4\,\text{m}^2$ of leaf per $1\,\text{m}^2$ soil surface. Assuming 20% of the leaf area is infected, a total of 100 million spores may be discharged from each square metre of soil area per day. These new spores are available to infect new plants.

Once airborne, these spores may travel long distances before falling to earth, though they may, of course, travel only a few centimetres before impingeing on a new leaf surface. When conditions are favourable to the pathogen very large epidemics indeed can occur.

Wheat rust. Perhaps the best studied outbreak was that which occurred in the U.S.A. in the spring of 1935. The outbreak was of wheat rust, an infection due to *Puccinia graminis*, a spore-forming fungus. The fungus survived the winter in a very small area of Texas and there was little rust on the wheat in early spring. However, spring was late and May was very wet, more than twice the usual amount of rain fell in Texas and rust quickly developed and began to spread. Spores were blown northward and found favourable conditions in the late spring crops of Kansas and Nebraska. The conditions in the southern part of Minnesota and neighbouring states were also favourable to rust – cold wet weather in May and June had delayed the crop. The first half of July was hot but still very wet and the stands of wheat were heavy. Masses of airborne spores were blown into these stands from Kansas and Nebraska. Rust developed quickly and before the end of the first week in July the epidemic was in full swing. Minnesota, North and South Dakota alone lost 45 million cubic metres of wheat.

Rust in coffee. A recent example of long distance airborne spread is that of the rust *Hemileia vastatrix* which attacks coffee crops. In the 1870s the rust spread through the coffee plantations of India and Sri Lanka wiping out the coffee plants, forcing the growers to turn to the cultivation of tea and changing the habits of the British who, until 1870 had been a nation of coffee drinkers. The fungus spread from India to Africa where it slowly extended its range until, by the late 1960s, it was well established in West Africa in the coffee plantations of Angola. In 1970 the fungus was found to have spread to Brazil where the crops were all of *Coffea arabica*, the most susceptible species. It seems probable that the spores were spread from the west coast of Africa, perhaps from Angola, by the Trade Winds which blow across the southern Atlantic ocean. The first move by the Brazilian government in an effort to keep the disease at bay was the destruction of all coffee bushes in an area approximately 50 km wide and 800 km long from Rio de Janeiro to beyond Bello Horizonte. This cost the government the equivalent of £5 million. The long-term answer is to breed new varieties of resistant crop plant, the short-term answer to spray the crops with copper fungicides. Both methods are expensive. Leaf rust disease, drought, frost and exhaustion of the soil, cut back the production of coffee to such an extent that the Coffee Institute of Brazil spent about £170 million in 1972 to plant 120 million new coffee trees and raise a further 150 million seedlings. In February 1972 the Brazilian Government announced that it would spend £440 million on improving coffee production. Rust diseases thus contributed to the high cost of coffee in the middle 1970s.

5.2.2 Water-borne transmission of fungal parasites

Water is another means of spore dissemination. But as a means of long distance transport it is not as efficient as wind dispersal. The spores may be carried along streams or rivers, but there is then little likelihood that they will reach a suitable host.

The probability of further infection is, of course, increased if the water is used for irrigation. Water is necessary to separate mucilaginous spores.

5.2.3 Vector transmission of fungal parasites

Some fungal parasites have developed spores that are carried around principally by insects. Most of the rust fungi produce, at one stage in their life cycle, pycniospores that function as gametes. When producing the gametes, the fungus induces a yellow colour in the host tissue and a brightly coloured sweet fluid is exuded. The bright colour and odour of the fluid attracts flies which inadvertently transport the pycniospores from one pycnium pouch to another, ensuring fertilization of the rust fungus. Similarly, *Claviceps purpurea* exudes a foul-smelling sticky solution which apparently contains up to 2 molar sucrose when producing pycnospores. The variety of insects which visit the host plant to feed on the pollen and sweet secretions become externally contaminated with spores, or they ingest the spores which are later excreted on a healthy plant. The genus *Ceratocystis* which causes Dutch elm disease, produces ascopores in fruit bodies called perithecia. The ascospores are not shot out of the perithecia but are exuded with a wet sticky matter which is then picked up on the hairy cuticle of beetles which inhabit the host trees.

Dutch elm disease. Dutch elm disease caused by *Ceratocystis ulmi* was first described in Western Europe in 1918 and reached Britain in 1927. Disease severity reached a peak between 1931 and 1937, and then declined. The disease spread from Europe to the U.S.A. in about 1930, resulting in continuous heavy losses, and has progressed across the continent. Although the disease has been endemic in Britain for more than 50 years, it has recently greatly increased in southern England. In 1972 the numbers of trees dying in the worst affected counties were:

Essex	109 000	Suffolk	47 000
Gloucestershire	98 000	Hampshire	42 000
Worcestershire	77 000	Herefordshire	26 000
Kent	66 000	Sussex	21 000

By 1977 the Forestry Commission estimated that about 11 million elms, or about half the population, had died in southern England.

PATHOLOGY. To establish itself, the fungus must find a wound in the bark and almost invariably these wounds are made by species of the bark-boring beetle *Scolytus*. The beetles chew into the bark to obtain the sap which flows in the phloem tubes. The fungus spores carried on the legs of the beetles are deposited in the wounds where they germinate and penetrate into the sapwood. Once in the woodcells, the fungus buds like yeast to produce masses of spores, these travel along the xylem vessels which convey water and mineral salts to the leaves. It is the toxin released from these spores which causes the tree to form plugs in the xylem vessels so that the upper branches and leaves wilt and die. The disease progresses rapidly and very young elms die within two weeks of becoming infected.

The life of the fungus does not, however, cease with the death of the tree. *Ceratocystis ulmi* is an extremely efficient parasite for once it has killed its victim the fungus changes its role and becomes a saprophyte. It then continues to live on the dead tissue of the host. At about this time the beetles, which have been continually boring galleries under the bark of the tree, lay their eggs. After the elm has died, young beetles emerge from pupae ready to take flight and transport countless fungal spores to other, healthy elm trees. There is, therefore, a close relationship between fungus and vector here.

During the severe 1972 outbreak in southern England, isolates of the fungus were collected from infected trees in the epidemic outbreak regions and compared with another group of isolates obtained from areas where the disease was only endemic. Examination of these isolates led to the belief that an unusually aggressive strain of *C. ulmi* existed in Britain and that this was responsible for the current severity of the disease. There were morphological differences between the aggressive and non-aggressive isolates. The aggressive strain was fast growing with aerial mycelium and was termed 'fluffy', while the non-aggressive strain was slow growing and had a waxy yeast-like appearance. Both strains were present in the outbreak areas but the aggressive strain predominated. The sudden appearance of this new strain in Britain was at first a mystery. It might have arisen by genetic change within an endemic non-aggressive gene pool. However, records indicated that the epidemic became evident in simultaneous outbreaks in two widely separated areas. Similar genetic change in both regions during the same period seemed highly unlikely. Therefore the epidemic strain must have been introduced into Britain from elsewhere.

In January 1973, rock elm logs recently imported from Toronto were examined at Southampton docks. Fungal mycelium was found in the fissures, together with live adults, eggs and larvae of *Hyalurgopimus rufipes*, the North American vector of Dutch elm disease. The *C. ulmi* strain from the Canadian logs closely resembled the current aggressive strain in Britain. We can conclude that the epidemic in Britain probably resulted from the importation of diseased logs carrying an aggressive North American strain of *C. ulmi*.

CONTROL. The relevance of strict quarantine laws regarding the importation of plants is clear from the story related above, but apart from these, other control measures need to be sought.

Pruning diseased elm portions has proven to be inefficient – it is impossible to be certain that all infected portions are removed and it does not prevent subsequent colonization of the remaining portion. The only sure method of keeping the fungus in check is to remove the dead elm and burn it down to the last twig. For many years investigators have sought chemical materials for the control of this disease and the recent development of several systemic fungicides has given new impetus to the search. One of these fungicides, Benomyl ((methyl-butyl carbomoyl)-2-benzimidazole carbonate), has been particularly effective in suppressing disease symptoms and eliminating the fungus from treated seedlings. The fungicide is readily absorbed from sand and translocated throughout the whole plant. Test trials with American elm seedlings largely favour Benomyl for future use against Dutch elm disease. Although they have some importance in the furniture trade, elms are used mainly as amenity trees and study of them and their diseases has therefore fallen outside the areas of horticultural crops, on the one hand, and the Forestry Commission, on the other, and so little research has been done. It is known that there are differences between the British elms in their susceptibility to Dutch elm disease, for example the hybrid *Ulmus minor* × *U. glabra* is more resistant than either of the parent strains; even within *U. minor* there are infra-specific differences. It may prove possible to select satisfactory resistant varieties, but of course any breeding programme of a tree must necessarily take a very long time.

5.3 Viral plant diseases

The total number of plant viruses is not known, but there is little doubt that crop plants in particular are susceptible to a great many, and some of the resulting diseases are economically important, for example barley yellow dwarf virus, potato virus X, tobacco mosaic virus.

The commonest effect of virus infection is the decrease in growth rate of the plant; so disease-ridden plants appear smaller and somewhat stunted compared with their healthy counterparts. Any part of the plant – roots, stems, leaves, flowers or fruits – may be involved, but usually the leaves provide the clearest signs of infection. Colour changes are the most common symptoms and, instead of being uniformly green, the leaves may exhibit spots, blotches, streaks or rings of differing colour; this is known as variegation (Fig. 5-1). Virus-induced variegation is best known in tulips; the self-coloured blooms characteristic of plants raised from seed, become streaked with other colours after they become virus-infected. This 'breaking' in colour occurs in several other species, for example chrysanthemum, carnation, gladiolus and sweet pea, but only in the tulip has great commercial use been made of this type of viral infection.

Fig. 5-1 Virus-infected leaf. The light patches are discoloured yellow in the original, this is the 'cytopathic effect'. Distortion of the leaf can also be seen.

Viruses are economically important only in plants in which they cause systemic infection, that is, where they can spread successfully from the initial point of entry to all other parts of the host. Most viruses exist as a variety of strains and have an extensive host range. But there are enormous differences between the behaviour of different viruses in one species of plant and of one virus in different species of plants, for example tobacco mosaic virus affects tobacco, tomato, and thorn apple and produces different symptoms in each.

5.3.1 Transmission

Unlike the fungal plant parasites, plant viruses lack any dispersal mechanism and cannot be dispersed by wind or water. So that, in the majority of cases, vector transmission is required to ensure spread of the virus. Those viruses not associated with a vector may be transmitted via the seed or pollen, by vegetative propagation, grafting or fungi.

Insects. Not all insects that feed first on a diseased plant and then on a healthy one are capable of transmitting virus. Since virus has to be introduced into a living cell before infection occurs, insects with piercing and sucking mouthparts are best suited as virus vectors. Aphids and leaf-hoppers are responsible for the transmission of a large number of plant viruses. One species of aphid alone, *Myzus persicae*, is the vector of up to 70 different viruses. Aphid-transmissable viruses can be divided into three main groups: stylet-borne, circulative and propagative. Stylet-borne viruses adhere to the stylets of the insects after penetration of a virus-infected plant. Viruses transferred in this way are usually of the filamentous type, they are picked up almost immediately after feeding and transmitted to a healthy susceptible plant. Circulative viruses (e.g. potato leaf roll virus, radish yellows virus) are swallowed and are transported, via the gut wall and blood, back to the salivary glands before the insect becomes infective. Propagative viruses are able to multiply in their vectors. All stylet-borne viruses are non-persistent, i.e. the virus is lost rapidly after its acquisition. Circulative and propagative viruses are termed persistent and the virus may survive in the vector for weeks or even months after acquisition from a diseased plant. All known viruses transmitted by leaf-hoppers, white flies, bugs and most biting insects are of the persistent type.

Fungi. Fungal plant parasites themselves can be vectors of viral disease. *Olpidium* is responsible for transmitting lettuce big vein virus and tobacco necrosis virus. The fungus and virus are in close association in doubly infected plant material. The virus adheres externally on the zoospores produced by the fungus. Thus, when the spores are dispersed, the virus is also dispersed and when the particle reaches a susceptible plant a double infection again occurs.

Nematodes. Only in the last 18 years have nematodes been shown to transmit plant viruses. Note was first taken when the crop yield in particular fields of strawberries was substantially lower than other fields. When the soil was examined in the bad growth zones, nematodes were found. They were absent in areas where crop growth was good. To verify that the nematode was the disease vector, healthy plants were grown in sterile soil and nematodes were subsequently added to the soil. As expected, the healthy plant soon went down with the viral disease. Scottish and English forms of raspberry ring spot virus are transmitted by the nematode *Longidorus*.

Seed. Only 10% of known plant viruses are seed-transmitted. It is not known why most viruses are unable to penetrate and infect the seeds of their host plant. The few examples of seed transmission are of virus that invades the embryo, but only in false stripe of barley has virus been demonstrated in all parts of the seed. Two viruses, namely those causing mosaic of bean (*Phaseolus vulgaricus*) and false stripe of barley, are pollen-transmitted and fertilizing the pistils of uninfected plants with pollen from an infected one will result in infected seed. Pollen transmission can be of considerable importance in woody perennials, an otherwise healthy tree may suddenly become infected when pollen lands on it.

Vegetative propagation and grafting. Any method of vegetative propagation, for example tubers, bulbs, corms, runners, suckers, rooting, cuttings or grafting, is likely to produce virus-infected progeny if the parent plant was infected. This usually occurs in plants which become systemically infected and such plants are capable of providing a continuous source of virus for as long as some vegetative parts of the plant remain viable.

Mechanical spread. Not many viruses are mechanically transmitted but this is by no means an unimportant mode of spread. Tobacco mosaic virus for example, which has a variety of hosts, often contaminates hands and clothing and can be passed onto healthy plants, for example during disbudding which involves handling. Viruses can also survive on farming equipment and be introduced into virus-free crops via tractors, ploughs, spraying machines etc. Animals, such as dogs, hares and rabbits, also aid in the dispersal of viruses.

§ 5.3 VIRAL PLANT DISEASES   67

5.3.2 Potato viruses X and Y (PVX and PVY)

One of the oldest records of virus diseases concerns the potato crop in some parts of Europe in the 1700s. There seemed to be a gradual degeneration of potatoes, the yield was decreasing each year because the tubers were becoming not only smaller, but fewer. The farmers called this annual lessening of crop yield 'running-out' or 'senility'. Unknown to them the degeneration was due to a virus infection spread from plant to plant by the insect inhabitants of the potato field. After a few years every tuber in the field was infected and each year when the farmer planted his stock potatoes, the disease was planted with them. Because this infection was due to a virus, isolation of the agent (PVY in this case), did not occur till the 1900s. Potato viruses prevalent at that time are still a problem in many parts of the world today. In Britain about £10 million is spent annually on the control of viral diseases of potato.

Potato virus X. PVX is one of the most prevalent of the many viruses infecting potatoes, it produces different diseases in different potato varieties. In most potato plants, PVX causes no local lesions, except perhaps a faint mottle. But, in some varieties, black spots develop on the leaves which, with the stem, then start to shrivel and die. In many cases the tubers are also affected, by the production of fissures, and are then unsaleable.

In the potato variety King Edward, infected plants develop necrosis of the growing points within 15 days, and usually die. Tubers from infected plants in the second year either fail to sprout completely or, paradoxically, go on to produce healthy plants. The Arran Crest and Epicure potato varieties exhibit a similar reaction to the King Edward, and infection usually leads to death of the plant. Usually plants in the second year of infection develop acute necrosis and eventually die.

CELLULAR EFFECT. Often in virus-infected tissue the cells develop cytoplasmic aggregations or inclusions. These inclusions are most frequent in the epidermal and hair cells and there is not usually more than one inclusion per cell. It is now agreed that inclusions found in virus-infected plants and animals consist largely of virus particles which have come together in numbers large enough to produce bodies visible under the light microscope. Intracellular inclusions are nearly always present in the mottled areas of infected potato plants and are known as X-bodies. X-bodies caused by the same virus in different hosts differ in appearance, as do X-bodies in the same host caused by different viruses. PVX produces a rounded inclusion in the potato varieties mentioned. The inclusions are vacuolated and are often situated close to, but not in direct contact with, the cell nucleus. Usually they are no bigger than twice the size of the nucleus.

Why this virus, or any plant virus for that matter, should settle out in the plant cells in the form of X-bodies is unknown. It may be that localizing the virus in the form of an inclusion might lessen its pathogenic effect, inclusions would then represent a host defence mechanism. However, this is still open to speculation. Cellular inclusion bodies are also encountered in many of the viral diseases of animals, for example Negri bodies of the rabies virus.

TRANSMISSION. PVX is one of the few viruses easily transmitted by mechanical means. Field workers, contaminated equipment, dogs and rabbits all aid in the spreading of virus. There is no true insect vector, but spread by grasshoppers is known to occur. The fungus *Synchytrium endobioticum* has been implicated as a vector and is able to transmit PVX inside the reproductive zoospores discharged from asexual zoosporangia.

CONTROL. This largely consists of starting with a crop of virus-free potatoes and preventing any mechanical spread of the virus. Virus-free plants can be achieved by propagating from the stem apices. Many viruses are unable to enter apical meristems and these uninfected tissue regions can be cultured aseptically on nutrient medium to produce virus-free plantlets. But great difficulty lies in excising a piece of tissue that is both virus-free and large enough to develop into a plant. Another method of obtaining virus-free stocks that has found wide application is heat therapy. In this method plants grown at high temperatures are cured of infecting virus. The growing potato plants are kept at 37°C in a moist atmosphere for 10 to 20 days after which time the virus is 'diluted out' from the plant. But by far, the best method of virus control and the cheapest is the breeding of virus-resistant varieties, for example the seedling U.S.D.A. 41956 was bred in response to the extensive infection of most potato varieties in the U.S.A. with PVX.

Potato virus Y. This virus too is a flexible rod with dimensions 730 nm × 10.5 nm. Besides potato, it causes disease symptoms in tobacco, woody nightshade, black nightshade, tomato, henbane, petunia and dahlia. In tomato, the virus can reach a weight of 1.5 g per litre infective sap. In the majority of potato varieties, leaf-drop streak symptoms or acropetal necrosis is produced.

In the President potato, disease symptoms develop about 3 or 4 weeks after infection. A blotchy mottling is produced affecting the leaves at the top of the plant, which eventually wither but remain hanging as if attached to the stem by a thread. In subsequent years, infected plants are small and stunted with very brittle leaves and stems. The leaves are twisted and bunched together so that the whole plant appears dwarfed and rosetted. Similar symptoms occur in Arran Banner, Up-to-Date, King George, Majestic and British Queen potatoes.

CELLULAR EFFECT. With some, but not all strains of PVY, intranuclear crystals and cytoplasmic membranous inclusions have been found. The crystal formation is associated with the cell nucleolus. When partially dissolved these isometric crystals yield many of the characteristic rod-shaped virus particles. The cytoplasmic inclusions are composed of the dictyosome of the Golgi apparatus.

TRANSMISSION. PVY is easily spread by mechanical means but there are also various species of aphids which are capable of transmission, for example *Myzus certus, M. ornatus, Aphis ramni*. As already mentioned, most filamentous viruses are transported on the insect stylets and PVY is no exception. All the above aphids carry the virus on the stylet tips. In addition, PVY is also transmitted by the spider mite *Tetranychus telarius* and, as with the aphids, the virus is stylet-borne.

CONTROL. Control of PVY is essentially the same as that described for PVX: recapping, that is the maintenance of virus-free potato stock by breeding resistant varieties, apical meristem propagation, heat curing, and prevention of mechanical transmission. Also, since PVY is vector-transmitted, vector control is essential.

Vector control can be achieved by isolation of plants. When susceptible plants are separated from each other by immune plants, virus spread is considerably lessened, especially if the intervening plants are suitable hosts for the vectors. Weeds and other plants are frequently reservoirs of viruses which are carried into crop plants from aphids, but a relatively short distance of isolation can decrease this danger considerably and complete removal of such weeds eliminates the hazard. When arthropods spread the virus successively from old to new crops, control has sometimes been achieved by omitting a crop from the cycle. Crops such as tomatoes can be grown protectively under glass to escape from virus-vectors but this is inconvenient for field crops.

The best method of vector control, and subsequently virus control, is the elimination of the vectors with chemical agents or insecticides. Insecticidal spraying is common practice now; pesticides used include phorate or demeton, methyl parathion, malathion and disulfoton. Systemic insecticides such as menazon are much more effective than contact ones because they penetrate into the tissues as the plants grow. But pesticides, although extremely valuable in disease control, must be used as sparingly as possible due to the long-term effects on other organisms including humans and, of course, the possible development of resistance in the insect pests.

PVX and PVY infection. When PVX and PVY together infect a plant, a new disease – rugose mosaic is produced. This disease commonly occurs in the U.S.A. since PVX is prevalent in the potato stocks there. Plants affected appear dwarfed and the tubers are reduced in size. The lower leaves have black necrotic veins while the upper leaves develop mottled chlorotic spots. Rugose mosaic was widespread in England in 1960. Some attacks were heavy, for example in Kent a 12-hectare crop of Arran Comet was 95% affected even though the 'seed' was only once grown and plants had been sprayed several times during the growing season. In the same year several potato crops in Berkshire and Hampshire had 30–50% of plants with severe mosaic and 99% of the Majestic crop in Oxfordshire was affected.

5.3.3 Tobacco mosaic virus (TMV)

Tobacco mosaic virus is one of the most infectious of the plant viruses. The virus particle itself is similar to PVY and PVX in that it is rod-shaped with dimensions 300 nm × 15 nm. However, the rod is not flexible but rigid.

TMV has an extremely diverse host range and has been found in tobacco, tomato, spinach, buckwheat, grapevine, blackcurrants, potatoes, orchids and apple and pear trees. As a result of its wide host range, TMV causes many diseases and only its effect in an economically important crop plant – the tomato – will be considered here.

Mild mosaic is a well-known disease in the tomato. The disease is due to infection by TMV and diseased plants appear mottled with raised dark green areas and distorted young leaves. In the summer, during conditions of high temperatures and light intensity, this mottling is quite severe. During the winter very little mottling is produced, but there is considerable stunting of growth and leaf distortion with the development of 'fern leaves'. Infection of the fruits just before ripening results in the tomatoes turning brown internally.

A particular strain of TMV – tomato aucuba mosaic virus – has been extensively studied due to its formation of pronounced intracellular inclusions. The aucuba virus causes the crown leaves of the tomato plant to crinkle or corrugate. In extreme cases almost the whole surface of the leaves turn pale-yellow to white with patches of raised green blisters. The fruit is usually unaffected but may sometimes become mottled. Tomato streak virus too is another strain of TMV. The disease this virus causes in tomato is a serious one. As the name suggests, the virus causes dark elongated streaks on the plant stems. The leaves show necrotic spots and subsequently shrivelled white sunken blotches occur on the fruit rendering them totally unmarketable.

In 1957–1961 TMV was severe and widespread in England. Many glass houses in the south-east were seriously affected, some of them completely. An unusually severe strain of TMV was apparent in 1961. It gave striking yellow symptoms in the Yorkshire and Lancashire crops and, besides the eventual bronze colouration of older leaves, green streaking of the stems was noticed. The more severe form of mosaic streak also caused some considerable losses in the same period. In addition, a number of nurseries developed mixed virus streak. This disease was due to infection with TMV+PVX. In two tomato crops in the south of England in which potato plants were present, mixed virus streak only occurred where the grower had tried to remove them. This emphasizes the particular importance of handling of the plants as a factor in spread.

CELLULAR EFFECT. The formation of intracellular inclusions is a common feature of diseases caused by tomato aucuba mosaic virus. Infected plant cells also often contain three-dimensional crystals; spike or spindle-shaped paracrystals; angled layer aggregates (small plates aligned and parallel virus particles).

TRANSMISSION. There is no known true insect vector of TMV, but grasshoppers can transfer the virus by mechanical contamination of the jaws. However, the virus is easily transmitted by mechanical means. TMV has also been thought to be seed-borne because young plants in sterile soil were nearly always infected. But infection in seedlings is probably due to virus residing on the seed coats and then entering the plantlet after germination.

CONTROL. Once again, because the virus is so infectious and easily transmitted, the maintenance of virus-free stocks and the prevention of mechanical transmission is essential to keep the pathogen at bay.

6 Postscript

A book of this length can only serve as an introduction to the very many ways in which microorganisms affect our lives. The contents represent an attempt to indicate the broad scope of the field, yet even so there are some major omissions.

For example, it has been necessary to leave out any indication of the intrinsic interest of microorganisms. Because microorganisms reproduce very rapidly, most advances in our knowledge of the way in which DNA and RNA code for, and transcribe genetic information, has come from a study of bacteria and their viruses. The latest steps in this field have given man the ability, in theory and to a great extent in practice, to insert specified genes into any chosen genetic system. This 'genetic manipulation' has aroused fears that a super-germ will be produced which will wipe out mankind either directly or indirectly via his crops. Yet an enormous potential for good also exists. In principle, plants such as wheat could be made to supply their own fertilizer by inserting nitrogen-fixing genes into the wheat genome. Microorganisms could be programmed to produce useful biochemicals such as the insulin needed by diabetics.

Even without custom-built compounds, there is an immense number of microbial compounds awaiting discovery and exploitation. More antibiotics exist than are currently in use or under study (it is usually the costs involved in testing these compounds to ensure safety which prohibits their free use). It is not impossible that conditions could be found in which even plastics could be broken down and recycled by microorganisms (indeed *Penicillium funiculosum* has already been recorded from plastics and may attack the plasticizer).

Microorganisms can be used to control other pests. Virus was deliberately spread amongst rabbits in Britain via arthropod vectors to cause the disease myxomatosis and reduce the rabbit population. Baculovirus was found to control spruce sawfly in Canada following an accidental spread of this virus in 1938. Commercial preparations of baculovirus are now available and have been used successfully in the U.S.A. to control corn earworm and in the U.K. to control pine sawfly.

In the past, microorganisms have been seen mainly as factors for evil; epidemics of disease have had far-reaching social and political consequences as well as causing misery to individuals by death and disease. At present the balance between good and evil is about equal;

there is still much decay and disease, but production of food and biochemicals and the harnessing of degradative enzymes are continually increasing. The future prospects are bright for the peaceful use of microbial metabolism.

Suggestions for Further Study

ANDREWES, C. H. (1967). *The Natural History of Viruses*. Weidenfeld and Nicolson, London.
BEVERIDGE, W. I. B. (1977). *Influenza. The Last Great Plague*. Heinemann, London.
BUSVINE, J. R. (1975). *Arthropod Vectors of Disease*. Studies in Biology no. 55. Edward Arnold, London.
CARTWRIGHT, F. F. (1972). *Disease and History*. Hart-Davis, London.
CLOUDSLEY-THOMPSON, J. L. (1976). *Insects and History*. Weidenfeld and Nicolson, London.
DEVERALL, B. (1969). *Fungal Parasitism*. Studies in Biology no. 17. Edward Arnold, London.
HAMMOND, S. M. and LAMBERT, P. A. (1978). *Antibiotics and Antimicrobial Action*. Studies in Biology no. 90. Edward Arnold, London.
HIGGINS, I. J. and BURNS, R. G. (1975). *The Chemistry and Microbiology of Pollution*. Academic Press, London.
HORNE, R. W. (1978). *Structure and Function of Viruses*. Studies in Biology no. 95. Edward Arnold, London.
HUGHES, W. H. (1974). *Alexander Fleming and Penicillin*. Priory Press, London.
KAPLAN, C. (Ed.) (1977). *Rabies. The Facts*. Oxford University Press, Oxford.
LONGMATE, N. (1966). *King Cholera. The Biography of a Disease*. Hamilton, London.
OLDS, R. J. (1975). *A Colour Atlas of Microbiology*. Wolfe Medical Books, London.
REID, R. (1974). *Microbes and Men*. BBC Publications, London.
ROSE, A. H. (Ed.) (1977). *Alcoholic Beverages*. Academic Press, London.
SMITH, K. M. (1972). *A Textbook of Plant Virus Diseases*. 3rd Edition. Longman, Edinburgh.
WHEELER, B. E. J. (1976). *Diseases in Crops*. Studies in Biology no. 64. Edward Arnold, London.
WHITNEY, P. J. (1976). *Microbial Plant Pathogenicity*. Hutchinson, London.

Slide/tape presentations on the impact of microbes in our lives prepared by the Microbiology in Schools Advisory Committee can be obtained from: Camera Talks Ltd., 31 North Row, London W1R 2EN. Topics include *Mushrooms, Penicillin, Viruses, Sewage, Food from Microbes* and *Methane*.